The Solution

to Society's Imbalance with Nature

Dayānanda dāsa

dayananda@cvc.guru

Centers for Vedic Culture

info@cvc.guru
https://cvc.guru/

Seeing the people intent on dominating her, the earth laughed and retorted, "Ha! These playthings in the hands of death think they can conquer me." -- *Bhāgavata Purāṇa* 12.3.1

- Acknowledgments i
- Preface ii

Introduction 1

- Terminology 2
- Vedic Cultural Ecology 8
- Branch of Science 11
- Metrics 14
- Two Cultures 18
- Overconsumption 22

Life, Reality, and *Dharma* 28

- Introduction 29
- Biological Life 32
- The Life Force 34
- Physical Reality 36
- Levels of Selfishness 40
- The Governing Source 43
- Transcendent Reality 51
- *Dharma* 53
- Time 58

Rejection of The Source 61
Comfort and Satisfaction 65
Modern Culture 67
Introduction 68
Science 69
Medicine 74
Rights 76
Democracy 79
Humanism 82
Socialism 86
Capitalism 89
Advancement 91
Religion 96
Entertainment 98
A Broken Machine 102
Vedic Culture 103
The Supreme Source 104
Four Barriers, Four Paths 108
Equality of Life 111
Pleasure 115

Acknowledgments

Although this book is only about one-hundred pages without the notes, it is the condensed result of decades of study and experience.

Foremost, I owe a tremendous debt for the education and guidance I received from my guru, Śrīla Prabhupāda.

After him, I am grateful for what I have learned from several teachers and professors in India, Iran, England, and the United States.

I am now fortunate to be part of a loose-knit team with Rūpa Vilāsa, Śarad-bihārī, Kṛṣṇa-kathā, Dānakelī, Mahāmāyā, and Bhūśaya. Without them, I would be lost.

Preface

This book presents the solution to modern society's imbalance with nature as given by the science of Vedic cultural ecology.

It is primarily for those who are concerned about the degradation of the environment and know little or nothing about Vedic culture.

Secondarily, the book is for the followers of the *Vedas* who wish to present their culture to outsiders.[1]

Vedic culture currently exists mostly in South Asia, but, since ancient times, it has spread globally.

Now, the number of people in Vedic societies around the world, not counting India, is more than the population of the U.K. About one billion live in India, and roughly 70 million are outside in many countries and continents.

There are at least two reasons to examine Vedic culture. First, considering that it exhibits an ancient yet still thriving, sophisticated culture, it is worth studying.

Second, Vedic culture is much more in balance with

nature than modern global society.[2] For that reason, this book presents Vedic culture as a living solution to the most significant environmental issues.

---

[1] Throughout the book, *Vedas* means the *Vedas* and their supplements like the *Upaniṣads, Bhagavad-gītā, Purāṇas, Itihāsas,* etc.

[2] One of the major purposes of this book is to explain how and why this statement is true.

Introduction

For info on Zoom meetings, email info@cvc.guru

Terminology

Here are summaries of some key terms, names, and concepts that are used and explained throughout the book.

Modern Global Culture

This is the term used to describe European or Western culture that has spread worldwide.

Ecosystem

An ecosystem is formed by the interaction of living organisms, including humans, with their environment.

Balance With Nature

Ecosystems, including the earth itself, attempt to maintain equilibrium. Thus, societies have a natural duty to keep themselves in balance with their ecosystems.

Imbalance With Nature

Modern global society has been overconsuming natural resources for decades, causing society to be out of balance with nature.

Overconsumption

Overconsumption causes significant damage to the environment and ultimately to humanity.

South Asian

When referring to the culture, instead of *Indian*, the preferred term here is South Asian.[1]

The *Vedas*

Vedic culture is based on the wisdom of the *Vedas*. Here, the word *Vedas* means the four *Vedas* as well as the *Upaniṣads, Bhagavad-gītā, Purāṇas,* and other supplementary literature.

Vedic Culture

Eight hundred years ago, Muslim invaders began calling the people of India *Hindu*, which means *Indian* in Arabic.[2] To them, Hindus included Muslims, Jews, Buddhists, and everyone else in India.[3]

Thus, Hindu is not a good name for the culture. Vedic culture is more accurate and more specific.

Vedic Cultural Ecology

This is the branch of science that analyzes the cultures of the world to determine whether they are strong or weak in maintaining a balance with nature.

Physical Reality

This is the perception of reality derived from seeing and experiencing the physical realm.

Life Essence or Life Force

The life essence is distinct from the physical. It is a non-physical energy that animates the body of each living being as its life force.

Causal or Governing Source

The *Vedas* explain that the universe has a governing source. That governing source unfolds, maintains, and withdraws the universe.[4]

Levels of Selfishness

The *Vedas* describe three levels of selfishness: (1) unselfish, characterized by maintenance and service; (2) selfish, characterized by ambition and creativity; (3) extremely selfish, characterized by ignorance and destruction.[5]

Human-centrism

Human-centrism means to selfishly think that humans are the central, most important part of existence or nature.

Religion

The ordinary meaning of religion is "the belief in and worship of a God or gods." However, it is an inadequate term when referring to Vedic culture.[6]

Dharma

Dharmas are sets of principles or laws used to govern everything from families to the universe.

Vedic Monotheism

Vedic monotheists define reality as the physical realm, the life force in all living beings, and the supreme controller of both. The controller is eternally distinct from the other two, which are His energies.[7] They never merge as one.

Viṣṇu and Kṛṣṇa

Viṣṇu is the governing or causal source of the universe, and He emanates from Kṛṣṇa.

Līlā

In Sanskrit, *līlā* (pronounced leela) means play, sport, pastime, or drama. It refers to the dramas Viṣṇu enacts when He descends to earth. Celebration of *līlās* is at the heart of Vedic culture, which might also be called *līlā* culture.

Vedic Pantheism

This doctrine considers that everything—the physical realm, the life force, and the controller—is ultimately one.[8]

Polytheism

This is reverence for many deities, generally presiding over the natural functions of the universe.

Atheism

This is faith only in physical reality and no identifiable source beyond it. Generally, atheists believe the physical realm and life are randomly generated.

---

[1] Bangladesh, Bhutan, India, the Maldives, Nepal, Pakistan, and Sri Lanka make up South Asia. Some scholars include Afghanistan, Tibet, and Myanmar (Burma).

When discussing the ancient Vedic culture that has transitioned through medieval into modern times, limiting that vast culture to a smaller modern country is not accurate. It is better to refer to a continent or subcontinent.

[2] Wikipedia, "Hinduism:" "In Arabic texts, al-Hind referred to the land beyond the Indus and, therefore, all the people in that land were Hindus."

Although the word Hindu has become widely accepted, it is not used in this book.

[3] Wikipedia, "Bahadur Shah Zafar:" "In 1857, after his defeat by the British, Shah Zafar composed a poem in which he referred to all Indians as Hindus. 'As long as there remains the scent of conviction in the hearts of Hindus (Indians), that long shall the sword of Hindustan (India) flash before the throne of London.'"

[4] The chapter "Governing Source" explains this term more fully.

[5] In Sanskrit, these levels are called *sattva*, *rajas*, and *tamas*.

[6] Wikipedia, "Religion:" "The concept of religion originated in the modern era in the West. Parallel concepts are not found in many current and past cultures. Scholars have found it difficult to develop a consistent definition, with some giving up on the possibility of a definition. Others argue that regardless of its definition, it is not appropriate to apply it to non-Western cultures."

[7] Vedic monotheism will be explained throughout. It is decidedly different from Christian, Jewish, or Islamic monotheism.

[8] Wikipedia, "Pantheism:" Advaita Vedānta considers Brahman alone to be reality and the physical universe is an illusory appearance of Brahman.

It is considered pantheistic.

Vedic Cultural Ecology

This book presents many of the findings of Vedic cultural ecology, which analyzes cultures and recommends solutions to environmental issues.[1]

The title of the book is *The Solution to Society's Imbalance with Nature.*

A key word is *imbalance.*

All species live in ecosystems. Within those systems, nature maintains a balance.[2] For various reasons, an ecosystem can change or die; however, living ecosystems seek to establish balance.

The earth is an ecosystem in which humans live.[3]

If societies and cultures do not find a balance with their ecosystems or with nature, they will suffer the reaction.

Therefore, solutions to environmental problems must bring human cultures into balance with nature. Recommending the means to achieve that is one of the goals of Vedic cultural ecology.

Here, we honestly say that Vedic cultural ecology is based on the perspective of the *Vedas* and Vedic culture.

We assert that it is dishonest for other texts on environmental science not to announce their cultural and ideological biases.[4]

The objectives and conclusions of environmental science are often biased toward modern societies, their politics, goals, and principles.

Moreover, despite the cooperation among branches of science, engineering, and politics, their environmental solutions are narrow and unproven on a large scale.[5]

Vedic cultural ecologists describe a working machine—Vedic culture.

In contrast, other environmental scientists report their view of a broken machine—modern global society—and they speculate on how it might be repaired.[6]

---

[1] *Cultural ecology* is the study of cultural adaptations to environments and how cultures affect nature. (Wikipedia, "Cultural ecology")

[2] "Ecosystem Resilience," *Earth and Planetary Sciences*, ScienceDirect.com. The natural tendency of an ecosystem is to remain close to its equilibrium state despite disturbances. That tendency is called *resilience*.

[3] "Ecosystem Earth," Science.org: "Although it is well accepted that Earth consists of many different ecosystems, human societies much less readily recognize that Earth itself is an ecosystem, dependent on

interacting species and consisting of finite resources."

[4] See, for example, "Biases in ecological research: attitudes of scientists and ways of control," *Scientific Reports* 11.226, nature.com.

[5] "Solving Environmental Problems," EstuaryMagazine.com: "Environmental problems can be complex and hard to resolve. The complexity arises because the components of the environment are linked, and their interactions may be separated by both time and distance."

Also, "Fixing the environment: when solutions become problems," by Marlowe Hood, phys.org: "In a world where climate change, air and water pollution, biodiversity loss, water scarcity, ozone depletion, and other environmental problems overlap, a fix in one arena can cause trouble in another."

[6] "Occam's Razor for the Planet," (GlobalEcoGuy.org) by Dr. Jonathan Foley:

"On the one side, there is a simple, clear, and effective solution that can help address environmental problems. It works. It could make a big difference. And it's ready to go today. But it may require a little adjustment on our part — perhaps consuming less stuff, wasting less, being more respectful of nature, or otherwise shifting our behavior.

"On the other side, there is a more complicated, technologically aggressive solution that is years or decades away from practical use. But it doesn't require us to change our ways. In fact, it usually allows us to consume *more*."

Vedic cultural ecology considers the latter solution—the complicated, technologically aggressive one—to be ultimately impossible to achieve. By allowing humans "to consume *more*," it will eventually cause humans to devour their environment and possibly themselves.

Branch of Science

Vedic cultural ecology is a branch of science. It uses methods and conclusions like any other social science such as sociology, anthropology, political science, etc.[1]

It also conforms with Vedic scientific proof, discussed below.

Methodology

Vedic cultural ecology analyzes current and historical cultures. Among them are South Asian, American, European, Russian, Chinese, Sub-Saharan African, Buddhist, Islamic, and others. The purpose of doing so is to examine how cultures exist in balance with nature.[2]

The analysis is particularly focused on doctrines, traditions, and social structures that reinforce balance. Those are based on perceptions of reality and society's laws. Thus, Vedic ecology is especially interested in those perceptions and sets of laws, called *dharmas*.

Qualitative and Quantitative

If the culture is still extant, it may be possible to use qualitative research by immersing in it, conducting interviews, or making direct observations.[3]

When observing existing cultures, quantitative methods are also effective, like comparing biocapacity and ecological footprint described below.[4]

Conclusions

The fundamental conclusion of Vedic ecology is that a society must predominantly accept a governing force in the universe that dictates a set of natural laws for society.

The governing force may be polytheistic, pantheistic, monotheistic, nontheistic, etc.[5]

Moreover, research indicates that a predominantly atheistic culture, whose nations do not follow the dictates of such laws, tends to overexploit nature.

One of the fundamental effects of universal laws is the control of greed that manifests as overconsumption of natural resources.

Vedic Science

Vedic cultural ecology conforms to the standards of

Vedic science.

One of the methods of proof in the *Vedas* is testimony, like that in a court of law. Such testimony may also be that of tradition or culture.[6]

In that context, Vedic cultural ecology uses the testimony of various cultures to arrive at the truth.

---

[1] Wikipedia, "Social science."

[2] Wikipedia, "Cultural ecology." Cultural ecology is the study of human interaction with the environment and the environment's impact on cultures.

Vedic cultural ecology is predominantly concerned with the effect humans have on their environment. Vedic ecology postulates that it is not enough for cultural ecologists to observe environmental issues and speculate how modern culture might respond. Instead, it is necessary to have a working model, in this case, Vedic culture, as a reference point.

[3] Wikipedia, "Qualitative research."

[4] Wikipedia, "Quantitative research."

[5] Buddhists, Jains, and Taoists are nontheists. They may not be concerned with a universal source like the pantheists and monotheists; however, they do accept governing forces like *karma*, *guṇa*, *dharma*, interdependent causation (*pratītya-samutpāda*), *tao*, etc.

In Eastern cultures, pantheistic, monotheistic, polytheistic, and nontheistic elements often combine in various ways, certainly far more so than in mainstream Christian and Islamic culture.

[6] This method is called *aitihya* in Sanskrit.

Metrics

This chapter is based on the terminology, data, and calculations of The Global Footprint Network.[1] There are two primary metrics:

First is the maximum ability of a country or the earth to produce resources and absorb waste sustainably. That is called biocapacity.

Second is the actual use and waste of a country or the earth, which is called its ecological footprint.[2]

The reason for these metrics is to caution humans not to exceed the capabilities of their environment. In other words, their footprint should not exceed biocapacity.

Biocapacity

Biocapacity measures the ability of an area, like a country or the earth, to produce natural resources and absorb waste and pollution.

The basic unit is a hectare.[3] Currently, on average, the earth has a biocapacity of 1.6 hectares per person.[4]

That means each person on earth has 1.6 hectares

for all their needs, comforts, waste, and pollution.

Those 1.6 hectares multiplied by the world population are available for sustainable resource production and waste absorption for all humans.[5]

Some countries have more hectares available for their citizens and some less, but the average for the earth's population is 1.6.

An example of a country that has a large biocapacity is the U.S., which has 3.7 hectares available for each person to use, waste, and pollute sustainably.

The U.K. has a biocapacity of 1.1 hectares per person. Russia has a biocapacity of 7.5. Saudi Arabia has 0.7.

Thus, the amount available for each person to use varies among countries due to population, climate, arable land, water, etc.

Ecological Footprint

The ecological footprint uses the same global hectare standard used by biocapacity.

On average, a citizen of the earth should have an ecological footprint of 1.6 hectares to properly maintain earth's population. That is because, as mentioned above, the average world biocapacity is 1.6.[6]

However, if measured by country, Americans may have an ecological footprint of 3.7 hectares, British 1.1, and Saudis 0.7 per capita. That is their corresponding biocapacity.

Most people in the world use more than their natural allotment.

Americans use 7.6 hectares per person. As citizens of the earth, nature provides them 1.6, and as citizens of America, nature provides 3.7. Thus, Americans are bad citizens in both regards.

In Europe, the ecological footprint averages about 4.0 per citizen.

Today, those in modern Vedic society use about 1.2 hectares on average. That makes them good world citizens.

Vedic Metric

Biocapacity measures the maximum that nature has to offer.

In contrast to focusing on maximum capability with a corresponding footprint, the *Vedas* urge humans to strive for minimum use.[7] Thus, the Vedic metric is the absolute minimum possible use, not the maximum that nature provides.

In Vedic society, maintaining a reasonable ecological footprint is achieved through social structure and cultural tradition, as explained in the section on Vedic culture below.

---

[1] The Global Footprint Network analyzes vast amounts of data from many international and national sources. It determines the biocapacity and ecological footprints of over 200 countries. (footprintnetwork.org)

[2] Definitions of Biocapacity and Ecological Footprint can be found at FootprintNetwork.org.

[3] A hectare is 2.47 acres of land and water.

[4] In 2019, there were approximately 12.2 billion hectares of biologically productive land and water on Earth, which is about 1.6 global hectares per person when divided by the world's population of 7.7 billion.

[5] That is, 7.7 billion humans times 1.6 equals 12.2-billion-hectare world biocapacity.

[6] When one's ecological footprint is larger than biocapacity, one is overconsuming; he is not helping to maintain the earth; he is not keeping society in balance with nature; he is an irresponsible earth citizen.

Since 1969, the citizens of the earth, on average, have been using more than their share for proper balance with nature.

Moreover, on average, the citizens of America and Europe have been using more than their country's biocapacity long before the rest of the world.

[7] *Īśa Upaniṣad* 1: "Everything is controlled and owned by the universal controller (Īśa). You should accept only your quota and not endeavor to acquire more."

Two Cultures

Americans and most Europeans use and waste twice as much as their countries' resources can sustain and absorb.[1] That means they would have to cut in half their environmental footprints to live sustainably with nature.

In other words, households, farms, industries, and the government would divide in half their use of food, water, electricity, gasoline, clothing, living space, entertainment, urban and suburban development, agricultural produce, government projects, the military, domestic products, and industrial materials.

They would generate half of the carbon waste, toxic chemicals, fertilizers, and other pollutants. They would discard half of the current refuse and leftover produce.

That is roughly the meaning of half.

In contrast to Americans and Europeans, the people of modern Vedic society already have a significantly reduced environmental footprint.

Consider a scenario where the entire populations of

Vedic society and America exchange countries.

In that case, in America, where one person lives, three people from modern Vedic society would live.

Despite such a large population increase, they would sustain themselves perfectly in balance with nature. That is because, compared to Americans, their consumption and waste is one-sixth.

The same thing could be done in Europe.[2]

The conclusion is that Americans and Europeans, who are both overexploiting and destroying the environment must change their behavior. That can be done through cultural change.

Managing Overexploitation

The *Vedas* declare: "One may lawfully claim what one needs to maintain oneself, but taking more makes one a thief who deserves nature's punishment."[3]

This injunction is one of many that are meant to control the natural desire of humans to overexploit.

On one end of modern Vedic society's spectrum, most people strive to follow this dictum to various degrees. On the other end of the spectrum are some who are greedy and exploit nature.

However, because the dominant themes of the culture are self-control, self-satisfaction, and reverence for nature and its source, most of the population live happily in balance with their ecosystems.

They highly respect those who live exemplary lives of minimal consumption. And the majority, who practice restraint, more than cancel the overexploitation of the few.

Modern Vedic Culture

Beginning after WWII, Vedic society has become increasingly impacted by modern, nature-destructive culture. As a result, more of its people have become exploiters of nature.

That is especially true since the development of global markets in the 1990s.

Nevertheless, modern Vedic society's environmental footprint remains much less than global society's.[4] The traditional culture is intact, and modern society should use it as a model to achieve balance with nature.

Even more can be learned about balance with nature from pre-WWII Vedic society and more still from pre-colonial society.[5]

[1] This is based on the Biocapacity and Ecological Footprint World Map, data.footprint.org.

The statement is "most Europeans" because Scandinavians are excluded. They do not use less than other Europeans. They use more; however, their large biocapacity can accommodate their high use.

[2] These scenarios are based on data from the Global Footprint Network.

[3] *Bhāgavata Purāṇa* 7.14.8.

[4] See the previous chapter and refer to the data on Global Footprint Network (footprintnetwork.org).

[5] "How Colonialism Spawned and Continues to Exacerbate the Climate Crisis," by Anuradha Varanasi, *State of the Planet, News from the Columbia Climate School*, news.climate.columbia.edu.

Also, "Environmental History and Historiography on South Asia: Context and some Recent Publications," by Michael Mann, *South Asia Chronicle* 3/2013, edoc.hu-berlin.de.

Overconsumption

The *Vedas* state that "Hunger, thirst, and anger can be satisfied, but greed can never be satiated even by conquering the world and enjoying everything in it."[1]

This concept establishes a fundamental premise of Vedic cultural ecology.

Currently, most scientists, politicians, business-people, and the public want to solve environmental issues through politics and technology.

However, the above maxim implies that those methods will not work. That is because the root problem is overconsumption.[2]

Being addicted to consumption is the same thing as the greed mentioned in the maxim.[3]

Greed

In moral terms, overconsumption means greed.[4]

In technical terms, overconsumption means having an ecological footprint that is too large.

It is so much consumption that nature cannot or

will not sustain it.[5]

The prognosis of Vedic cultural ecology is that overconsumption is now a cultural addiction that will continue to increase.[6] Modern society cannot control its addiction.[7]

Thus, society must receive proper guidance.

Addiction

There are proven methods for curing addiction. They begin by receiving help from experts, who often suggest changing one's environment, effectively one's personal culture.

Another important aspect of curing addiction is addressing pleasure. Experts encourage addicts to seek constructive ways to enjoy life.

Here is a discussion of those methods.

External Help

Addicts are typically urged to get external help from a physician, treatment center, therapist, continuing care, peer-support, etc.[8]

One cannot cure oneself; it is nearly impossible. A wise person knows when to seek good counsel. Ignorant people tend to be arrogant, thinking that they can solve all the challenges of their lives.

Thus, Vedic cultural ecology strongly suggests that modern addictive culture should seek treatment from the expert Vedic culture.

Modern culture is diseased; Vedic culture offers the cure.

Changing Environment[9]

Nearly everywhere, people are encouraged to consume as much as possible; most people overconsume.

Modern society must change its environment from one of consuming nature to living in balance with it. It is a big task. Politics, laws, and technologies play only a small part.

Changing Association

Along with changing one's environment, one must keep company with sober friends and associates and give up the addicted ones.[10]

A major theme in Vedic culture is association with those who live simply and have minimal possessions.[11] That principle of association is powerful and pervasive in Vedic society; it is effective.

Changing Pleasure

This is the principle most emphasized by Vedic cultural ecology, which asserts that Vedic society has the kind of enjoyment that can control addiction to overconsumption.

There have been many studies on pleasure and addiction.

For example, in her book, Suzette Glasner-Edwards writes, "People with the most success in staying sober tend to get involved in a range of pleasurable activities and do them frequently.

"These activities can replace the time and energy that they had been spending on addictive behaviors, enabling them to experience pleasure without the devastating consequences of alcohol or drug use."[12]

Vedic cultural ecology prescribes a similar solution to society's addiction. The pleasure derived from overexploitation must be replaced by pleasure that does not destroy nature. The means to do so is presented in the chapter entitled "Pleasure."

---

[1] *Bhāgavata Purāṇa*, 7.15.20.

[2] "Earth Overshoot Day: The Bitter Truth of Overconsumption in the

Global North," Cody Peluso, 2023, PopulationMedia.org: "From excessive energy consumption to the production and disposal of vast amounts of goods, the Global North's consumption habits place immense strain on ecosystems and contribute to greenhouse gas emissions, deforestation, and biodiversity loss."

[3] "Overconsumption and the environment: should we all stop shopping?" TheGuardian.com: "Over-consumption is at the root of the planet's environmental crisis."

[4] "The Role of Consciousness in Overcoming Greed – Humanity's Tool for Combatting the Environmental Crisis," by Sophie Raymond, Queen's University, queensu.ca: "Many of the environmental issues plaguing the Earth today can be traced back to greed. Greed manipulates us to the point that we consume more than we need, and the materialistic desires of most people in the Western world require more and more to be satisfied. Greed is even identified as one of the seven deadly sins in Christian teachings."

[5] "How Overconsumption Affects the Environment and Health, Explained," by Seth Millstein, sentientmedia.org: "Of all the human practices that are gradually destroying the environment, overconsumption is one of the most significant and least discussed. We are using up our planet's resources faster than they can regenerate."

[6] "Addiction = Greed = Selfishness," *Choice Recovery*, choiceiop.com.

[7] "Overconsumption: The Silent Addiction Impacting Our Brains and Planet," Medium.com, by Iswarya: "There's an addiction that seems to go unnoticed and even encouraged in our society: the addiction of overconsumption."

[8] "How to Break an Addiction: A Guide to Overcoming Addiction" by Stacy Mosel, L.M.S.W., American Addiction Centers, americanaddictioncenters.org.

[9] "Environmental Risk Factors for Developing an Addiction," Greenestone Centre for Recovery, greenestone.net.

[10] "5 action steps for quitting an addiction," Harvard Health Publishing, Harvard Medical School, health.harvard.edu.

[11] *Śrī Caitanya-caritāmṛta Madhya-līlā* 22.54: "Even a moment's association with those who have no possessions brings all success." (*akiñcana bhakta*)

[12] *The Addiction Recovery Skills Workbook: Changing Addictive Behaviors Using CBT, Mindfulness, and Motivational Interviewing Techniques*, by Suzette Glasner-Edwards and Richard A. Rawson.

Life, Reality, and *Dharma*

For info on Zoom meetings, email info@cvc.guru.

Introduction

The two sections following this one analyze the ways modern and Vedic cultures treat nature.

Before that analysis, it is necessary to understand how people in those cultures perceive life and the world around them. In addition, it is important to examine the sets of laws that govern people in those cultures.

Those perceptions and sets of laws, called *dharmas,* shape a society's attitudes and behaviors toward nature.

This section begins by examining the definition of life and its purpose.

The next discussion is about perceptions of reality dominant in the two cultures. Those perceptions are influenced by concepts of the source of the universe, time, and other factors.

Near the end of the section is *dharma,* which is a theme throughout the book.

The following summaries are expanded in the succeeding chapters:

Life

In modern science and education, life is generally considered to be physical, meaning it is made of chemicals.

The Vedic perception is different. The *Vedas* state that the life force animates the physical body and is a separate energy.

Understanding life and its purpose is an essential part of society's attitude toward nature.

Reality

The dominant perception of reality in modern culture is physical. People are educated to believe the observable universe is reality. At least, the idea is that it is a sufficient perception with no need to go further.

In contrast, people in Vedic culture are oriented around a transcendent reality that lies beyond the physical.

A society's predominant perception of reality influences its attitudes and behaviors toward its environment.

The Source of the Universe

An essential consideration when discussing society's relations with nature is whether it is human-centric or source-centric.

In human-centric society, people are principally concerned with humanity, not necessarily nature.

In Vedic society, which is source-centric, humans revere nature and its laws because they were established by the source of the universe.

Dharmas

Dharmas, meaning sets of principles and laws, influence a society's behavior toward nature.

Biological Life

Based largely on the Theory of Evolution, biologists define life as the process of survival.[1] They have no proof, but nonetheless, that is generally their view.[2]

Trifonov opines, "Since there is no consensus for a definition of life, most current definitions in biology are descriptive. Life is considered a characteristic of something that preserves, furthers, or reinforces its existence in the given environment."[3]

In contrast, the *Vedas* declare that life is formed of existence, knowledge, and enjoyment.[4]

Trifonov mentions existence but not knowledge or enjoyment. Generally, the biological view is that knowledge and enjoyment are subordinate to existence or survival. Their root cause is chemicals.

The *Vedas* agree that knowledge and enjoyment may be achieved through the body's chemicals; however, they are not the root cause.[5]

Instead, it is the life force that acquires knowledge and experiences enjoyment through its mind and body—that is, through chemicals.[6]

[1] "Survival of the fittest," is a term made famous in *On the Origin of Species* by Charles Darwin, which suggested that organisms best adjusted to their environment are the most successful in surviving and reproducing.

[2] From Wikipedia, "life:" "What is life? It's a Tricky, Often Confusing Question." By Chris McKay, *Astrobiology Magazine*: "The definition of life has long been a challenge for scientists and philosophers. This is partially because life is a process, not a substance."

McKay's assertion that life is a process not a substance is hypothetical. Scientists can only see life processes. They cannot see life, so they speculate.

[3] "Definition of Life: Navigation through Uncertainties," by Edward N. Trifonov, *Journal of Biomolecular Structure & Dynamics.*

[4] *Viṣṇu Purāṇa* 1.12.69: *hlādinī sandhinī samvit* (synonymous with *sat-cit-ānanda*, existence, knowledge, and enjoyment)

[5] *Bhagavad-gītā* 13.21: "Nature is the source of all physical causes and effects, whereas the life force is the root cause of suffering and enjoyment."

[6] It is possible for the life force to acquire knowledge and experience enjoyment that are not limited by the physical body. The *Vedas* teach the process of doing so.

The Life Force

Definition and Purpose

The *Vedas* explain that the life force inhabits the body-mind vehicle and animates it.[1]

Before the global influence of Christianity, most cultures had an animist concept of nature. Over time, with its less nature-friendly doctrine, it replaced most of the natural, animist cultures.[2]

The *Vedas* explain that the life force is the same in all beings despite the differences in physical forms.[3]

The variety of forms—from viruses to galaxies—have unique roles in the functions or maintenance of the universe. Biologists hypothesize that the purpose of life is survival; however, the *Vedas* declare that its higher purpose is the maintenance of nature.[4]

Since life's purpose is maintenance, humans do not have the right to interrupt the workings of nature by overexploiting it.[5]

That is the natural perspective on the role of life.

Identifying as Life Force

One's true self is the life force.[6]

The causal source, explained below, animates nature and the universe with life.[7] One's self or life force is an energy of the source and eternally connected with him.[8]

The life force and the causal source constitute a reality that transcends the physical realm.

---

1 *Bhagavad-gītā* 2.17, 18.61.

2 *The Bible,* Genesis 1.26-28.

3 *Bhagavad-gītā* 18.54.

4 *Bhagavad-gītā* 7.5: "One of nature's energies is physical and the other is life force. The life force in all beings maintains the animated universe."

5 *Bhagavad-gītā* 3.12.

6 *Bhagavad-gītā chapter* 13.

7 *Bhagavad-gītā* 7.6, 7.10, 14.3.

8 *Śvetāśvatara Upaniṣad* 6.7–8.

Physical Reality

Vedic cultural ecology defines *physical* reality as the realm described by empirical science.[1] That physical realm also includes mental and social perceptions of reality.

Identifying as the Physical Body

The predominance of physical identification is unhealthy. An individual or society that identifies with the life force is more balanced with nature.

Too much identification with the physical realm typically promotes selfishness and human-centrism.

It is often characterized by possessiveness and too much attachment to immediate goals and enjoyment.

Those with such a perception of reality typically think that they alone produce results without acknowledging nature or its source.[2]

Excessive physical identification also causes violence and war. The *Vedas* equate it with ignorance.[3]

They declare that "One who identifies with physical reality thinks of himself as his body and his extended self as his family and country. Such a perception is no better than that of a cow or a donkey."[4]

Māyā

Māyā means illusion.[5]

The concept of *māyā*, described in the *Vedas*, declares that physical reality, including mental, social, economic, and political realities, is illusory, like a dream.[6]

However, that does not give a person license to act whimsically or irresponsibly.[7] Instead, one should dispel the illusion of *māyā* by following the direction of *dharma*.[8]

The *Vedas*, which establish *dharma*, are beyond the veil of illusion.[9] Thus, the *Vedas* must verify what is real and what is not.[10] Otherwise, one's perception of reality will be clouded, and one's actions will be irresponsible, or worse, destructive.[11]

The Invisible Repository

According to the *Vedas*, "There is an invisible repository of elements that is eternal and contains nature's qualities and genesis. It is yet without variety. When it unfolds with all its varieties, it is called physical nature."[12]

The causal source, described below, injects time and life into the invisible, inanimate, formless, and indeterminate repository.

Doing so animates the universe with structure, qualities, causes and effects, movement, and life.

It is only possible to understand the invisible repository and its source through an internal process that goes beyond direct perception.[13]

---

[1] For simplicity, Vedic cultural ecology defines *physical* and *empirical* as synonymous although some scientists distinguish between the two.

[2] *Bhāgavata Purāṇa* 3.26.5, "When one's vision is limited to a dynamic nature, unfolding with myriad forms, one's perception is distorted. One thus misidentifies with physical reality. He thinks that he, as life force, is acting, but in truth, it is nature that is doing everything."

[3] *Bhagavad-gītā* 18.25 "The behavior of one too identified with his body who does not take proper guidance and who is violent is ignorant."

[4] *Bhāgavata Purāṇa* 10.84.13.

[5] *Vedānta-sūtras* 3.2 provides a discussion on *māyā*.

[6] *Bhāgavata Purāṇa* 4.12.15: "Dhruva realized that, like a dream manufactures fantasies, *māyā* creates this universe."

[7] *Bhāgavata Purāṇa* 10.40.23-26: "I am bewildered by *māyā*, and I think in terms of 'I' and 'mine.' I confuse the temporary with the eternal, the

body with the self. I am like a fool distracted by a mirage."

[8] *Bhāgavata Purāṇa* 1.2.9: "*Dharma* is meant for liberation (from *māyā*), not material gain, which puts one under the control of *māyā*."

It is possible to follow *dharma* to become materially enriched. However, that is a lower form of *dharma*. This *Bhāgavata* verse states that true *dharma* is meant to liberate one from the illusion of *māyā*.

[9] *Yajurveda, Ṣaḍviṃśa Brāhmaṇa*: "The *Vedas* are *apauruṣeya* (not originating from humans). They are considered the ultimate authority or proof of reality (*mūla pramāṇa*) in the *dharma-śāstras* by Baudhayana, Parāśara, Vedavyāsa, Gautama, Vaśiṣṭha, Āpastamba, Manu, and Yājñavalkya."

These are the ancient authors of *dharma-śāstras*. According to Vedic tradition, they are superior to modern humans, and in some cases, immensely so.

[10] *Bhagavad-gītā* 16.24: "The *Vedas* must be used to determine what to do or not do."

[11] One may argue that, since the world is like a dream, destructiveness does not matter. However, beyond the illusion is a reality. To achieve that reality, destructiveness is counterproductive. It thickens the cloud of ignorance.

[12] *Bhāgavata Purāṇa* 3.26.10-11.

[13] *Bhagavad-gītā* 13.35: "Those who, with the eyes of knowledge, see the difference between the physical and the life force, can understand the process of becoming liberated from the physical."

Levels of Selfishness

The *Vedas* state that the physical universe unfolds under the control of three aspects of nature.[1]

These three governing aspects are described variously; however, for simplicity, Vedic cultural ecology defines them as levels of selfishness.

Some combination of the three aspects influences everyone. In an individual or society, one aspect is typically predominant with the other two having less control.

Illumination[2]

This is the least selfish aspect; its main characteristics are service and maintenance.

It is liberating. It influences happiness, self-satisfaction, and the ability to discriminate between physical and transcendent reality.[3]

Dimness[4]

The next aspect is often called passion in the sense

of ambition, creativity, or intense endeavor.

It is characterized by depression, anxiety, attachment to physical reality, longing, and greed.

Darkness[5]

This is the lowest, most selfish quality. Its influence is madness, ignorance, illusion, violence, and destruction.

Social Classes

To gradually free oneself from selfishness and rise to the level of illumination, one must follow *dharma*, discussed later in this section.

In all societies, classes emerge wherein people with a predominance of one of the three levels of selfishness group together.[6]

In modern society, for the most part, the influential classes are dominated by selfish people. They are the wealthiest, most powerful, and most famous, and they are typically the most respected.

However, regardless of how normal that may seem, it is unnatural.

In a natural society where maintenance is the primary theme, there is an influential class of people who are unselfish or illuminated.[7]

In Vedic society, the illuminated class is the most honored and supported. They guide society toward service to humanity, nature, and the causal source of nature.

---

[1] *Bhagavad-gītā* 14.5: These three aspects are sometimes called qualities or modes of nature.

[2] *Bhagavad-gītā* 14.6. In Sanskrit, *sattva* is literally trueness, goodness, realness.

Bṛhad-āraṇyaka Upaniṣad 1.3.28: *tamasi mā jyotir gamaḥ*, "Do not go to *tamas* (darkness); go to the light, to illumination, which is Brahman."

Those who pursue such illumination are the *brāhmaṇas*, those influenced by *sattva*.

[3] *Bhāgavata Purāṇa* 10.33.32.

[4] *Bhagavad-gītā* 14.7. In Sanskrit, *rajas* is literally dimness, impurity, or dust. It also means passion in the sense of active and urgent.

[5] *Bhagavad-gītā* 14.8. In Sanskrit, *tamas* is literally darkness or ignorance.

[6] *Bhagavad-gītā* 18.41: "The classes and their functions in society emerge according to characteristics influenced by levels of selfishness."

This natural grouping is a general tendency. There are exceptions.

[7] *Bhāgavata Purāṇa* 7.1.10: "Maintenance is carried out through the first quality, creation through the second, and destruction through the third."

The Governing Source

One may disagree on whether the governing force of the physical realm is pantheistic, polytheistic, monotheistic, or a combination.

But an atheistic force is impossible. The universe is not controlled or determined randomly or probabilistically as some insist upon.

Such a hypothesis is invalid and cannot be proven. Humans cannot directly understand the source and controller of the universe. That is because human mental and mathematical abilities are limited to a subset of the universe. They cannot reach beyond their limitations.[1]

Although one may not be able to perceive the source directly, one can see the results of his *dharma*. In Vedic culture, people follow the *dharma* or laws established by the source, and they live in balance with nature.

In contrast, people in modern society violate those laws, overexploit nature, and cause environmental problems.

To an intelligent person, both situations prove that nature is structured by laws. When a majority follow them, society is a good citizen of nature. By violating them, an outlaw society is created.

Deterministic Universe

The *Vedas* explain that the supreme source unfolds a deterministic or governed system—the universe. It is deterministic because the source determines, governs, or controls it.

He does so indirectly and directly.

He governs it indirectly through his controlling agents as an executive power, like a president or king, manages through his representatives.[2]

And he directly controls the states and functions of the universe by spreading himself throughout as universal consciousness.[3]

A fundamental goal of many in Vedic society is to connect with the source by serving and contemplating that all-pervading aspect.[4]

Fatalism

A universe that is determined by a causal source implies fatalism.[5] In modern society, fatalism is anathema. People want to control their destiny.

Some believe fatalism means that humans have absolutely no control over their future. However, the *Vedas* declare that is the wrong idea.

Humans have the power of discernment,[6] which means they can control themselves by discerning what is *dharmic* and what is not.[7]

The *Vedas* explain that the mind and discernment are two separate forces.[8] The mind is helplessly carried away by its attraction to the physical realm.[9] Thus, it is under control of fate.

Cultivating discernment is the key to unlocking freedom, which is defined as overcoming the mind's bondage to the physical realm.[10]

At birth, one's discernment is covered by a greater or lesser degree of selfishness, making it harder or easier to discern between right and wrong behaviors.[11]

The best way to remove that covering is to stop identifying as a physical being and instead identify with the life force and the source.[12]

However, it is not a mental process; one cannot just switch identities by deciding to do so. Instead, the process to rid the cover of selfishness begins with following *dharma*.[13]

Those who do not follow *dharma* and try to create their destiny are carried away by *māyā*.[14] They are under the illusion that they are free to do what they like. However, they are always under the control of *māyā*.

Ironically, trying to control one's destiny puts one at odds with nature. That is irresponsible.[15] One can get the best results by responsibly following *dharma*.

Doing so unlocks true freedom, which is freedom from *māyā*.[16]

Unfolding the Universe

According to the *Vedas*, the causal source exists beyond the invisible repository that contains the unformed elements of the physical universe. The repository does not yet contain either time or life.

The causal source, or transcendent personality,[17] with the glance of time, causes the repository to unfurl and reveal the physical universe.[18] In other words, time is the feature of the source that activates the unfolding of the universe.

His glance also impregnates it with the life that animates everything.[19]

Vedic experts interpret that source as pantheistic, monotheistic, or a combination of the two.[20]

Vedic pantheists declare that the source encompasses nature, the life force in all beings, and the controller of both—all as one entity.[21]

Vedic monotheists do not accept that the source is one with the other two. Instead, they are his subordinate energies.[22]

Both pantheists and monotheists reject atheism and a random universe or random nature.

The *Vedas* describe the source as the controller, supreme, maintainer, destroyer, cause of all causes, and the most opulent.[23] That source is visibly reflected in the world as the essence or superlative aspect of everything.[24]

---

[1] *Brahma-saṁhitā* 5.34: "One cannot approach the source even traveling at the speed of mind for billions of years."

[2] *Bhāgavata Purāṇa* 10.9.19: "Kṛṣṇa controls the universe, including all the controllers within it."

Bhāgavata Purāṇa 3.29.38-44, summary: Viṣṇu is the time factor, the controller of controllers, the support of all. Due to fear of Him, His subordinate controllers, unfurl, maintain, and destroy the physical realm. Everything functions due to fear of Him.

[3] *Bhāgavata Purāṇa* 1.2.32: "The one source, as the universal self, pervades all things, like fire permeates wood."

Bhagavad-gītā 13.3: "I (the source) am the knower of all, the consciousness of the universe."

[4] *Muṇḍaka Upaniṣad* (3.1.2) and *Śvetāśvatara Upaniṣad* (4.7): "The individual self or life force and the universal life force are both separately within the body, but the life force is distracted by physical enjoyment and frustrated in his attempts. If he turns to his friend, the universal life force, he becomes free from suffering."

There are two understandings conveyed here.

First is that the life force is not truly connected with the physical realm. The life force falsely identifies with that realm. The body eats, has sex, and enjoys its family. But it is physical nature that is doing those things, not him. Yet the life force tries to enjoy them and gets frustrated in the process.

Second is that if the life force turns his face toward the local aspect of the source, who is his true friend and with whom he has an eternal relationship, he can be relieved from suffering. That is liberation.

Regarding the first point, the *Bhagavad-gītā* 3.27 explains, "The life force falsely thinks himself the doer and enjoyer of functions (like eating and sex). Those functions are truly performed by physical nature under the control of the universal life force."

Bhāgavata Purāṇa 11.28.6-7 confirms that the soul of the universe is the controller of physical nature.

In this book, soul of the universe, universal life force, local aspect of the source, and universal consciousness are synonymous. In Sanskrit, the term is Paramātmā, meaning Supreme Self or Soul, which is often translated as Supersoul.

[5] Wikipedia, "Fatalism." Fatalism is the concept of the universe as a de-

terministic system.

[6] Discernment is called *buddhi,* which can be used to overcome one's fate (*karma*) as a kind of free will. The mind is under the control of *karma,* so free will is not the power of the mind.

Also, one does not overcome fate (*karma*) solely by personal endeavor. It must be done through linking the *buddhi* with a higher source.

The process for doing so is to cleanse the *buddhi* from the contamination of selfishness (*kāma*). That cleansing is done through the discipline of following *dharma.*

[7] When families and societies are oriented around universal *dharma,* people receive proper guidance naturally. (*Bhagavad-gītā* 1.39-43)

It is not possible to cultivate proper discernment by oneself. Help is needed, typically by a *guru* or teacher and through proper social structures. (*Bhagavad-gītā* 4.34)

[8] *Bhagavad-gītā* 3.42.

[9] *Bhagavad-gītā* 2.60-63.

[10] *Bhagavad-gītā* 3.43; *Mahābhārata, Vana-pārva* (313.117) quoted in *Caitanya-caritāmṛta, Madhya-līlā* 17.186.

[11] *Bhagavad-gītā* 18.41.

[12] *Bhagavad-gītā* 3.43.

[13] *Bhāgavata Purāṇa* 11.17.1-7; *Bhagavad-gītā* 4.7-8; *Bhāgavata Purāṇa* 1.2.9; *Bhāgavata Purāṇa* 7.15.12.

[14] *Bhagavad-gītā* 7.15.

[15] *Bhagavad-gītā* 16.7, 16.9, 16.16.

[16] *Bhagavad-gītā* 7.14.

[17] The personality (*puruṣa*) is the animator of the universe.

This may be a difficult concept to grasp for Westerners who identify personality only with humans, dogs, cats, etc.

In the Vedic perspective, moving (animated) beings—planets, galaxies, trees, ants, etc.—have personalities associated with them. In other words, motion means that they are animated by life and have personal characteristics.

[18] *Bhāgavata Purāṇa* 11.24.20: "As long as the *puruṣa* glances upon nature, the physical world continues to exist, perpetually generating the great and variegated flow of universal creation."

[19] *Bhagavad-gītā* 7.5: "Beyond the physical universe is the life force that animates it."

Bhagavad-gītā 8.3: "The nature of the life force, the self, is the same as that of the indestructible pantheistic source."

[20] *Bhāgavata Purāṇa* 1.2.11: "The source, or highest reality, is described as pantheistic, omni-present, and monotheistic."

Regarding the polytheistic concept of the source, *Bhagavad-gītā* 10.2 explains "The supreme source is the origin of the gods (governors of nature)."

Bhagavad-gītā 9.23: "Those who sacrifice to the gods, the polytheists, are actually sacrificing to the supreme monotheistic source but in the incorrect way."

[21] *Śvetāśvatara Upaniṣad* 1.12.

[22] *Bhagavad-gītā* 14.3: "Pantheistic Brahman is the womb of nature that I, as the monotheistic source, impregnate, thus giving birth to all living beings."

[23] *Bhagavad-gītā* 11.36-39.

[24] *Bhagavad-gītā* 10.41: "All that is wonderful, powerful, and beautiful represents a spark of my splendor."

Transcendent Reality

The *Vedas* explain that transcendent reality is superior to the physical.[1]

Transcendent reality is comprised of the source, the life force in all living beings, and *māyā* or the physical realm. Although *māyā* is within it, transcendent reality is the true reality free from the illusion of *māyā*.

To identify with transcendent reality, one must be able to discern between the true reality and the illusory perception of reality (*māyā*).

Without such discernment, the physical reality remains one's identity. Such identity promotes selfishness, greed, and suffering.[2]

The process for transferring one's identity to transcendent reality is through supreme *dharma*. Everyone must follow the *dharma* that governs the physical universe.[3] However, supreme *dharma* enables one to connect with the source and achieve transcendent reality.[4]

[1] *Bhagavad-gītā* 7.5, 8.20-21; *Muṇḍaka Upaniṣad* (2.2.9-11)
Also, *paravyoma—Śrī Caitanya-caritāmṛta Madhya* 20.213.

[2] *Bhagavad-gītā* 2.62-63: "Identification with the physical results in a magnetic attraction to its objects. From that attraction arises desire, greed, and anger. Those cause one to become falsely allured and illusioned. From that state, one loses discrimination, becomes a slave to physical desires, and thus suffers."

[3] Not following universal *dharma* is like breaking the laws of the state. The result is imprisonment and suffering.

[4] *Bhāgavata Purāṇa* 1.2.6: "Supreme *dharma* is that by which humanity can achieve *bhakti* (direct service) to the supreme source."

Dharma

Levels of *Dharma*

Dharma is a set of principles or laws.

The *Vedas* explain that there are *dharmic* levels.[1] The highest *dharmas* are dictated by the governing source and given in the *Vedas*.[2]

A group or community might have a man-made *dharma*, which means a set of rules encouraging responsible actions and restricting harmful ones.

Then, on a higher level, a district, province, or state may have a *dharma*. That *dharma* overrides the lower one. This means all the rules in the lower *dharma* are valid provided they do not contravene the higher *dharma*.

An example is that a state law in the United States may not contravene a federal law.

The *Vedas* declare that *dharmic* levels progress to the *dharma* whose rules bring order to the universe.[3]

Humans have the right to establish laws to govern

themselves. However, those laws must not contravene nature's *dharma*, or else they will be overridden.

It is not merely a question of faith or belief in such a system. Levels of *dharma* are an observable phenomenon.

Regarding universal *dharma*, empirical scientists have observed rules involving time, energies, particles, gravity, etc. Those inflexible rules or laws are a part of universal *dharma*.

Thus, since the rules of *dharma* are everywhere, it is logical that humans must conform to a natural societal *dharma* that is superior to man-made *dharmas*. It is not only logical, but it is also responsible.

Society's *Dharma*

Among humans, the *dharma* that is dictated by nature and its governing source starts with responsibilities and restrictions for a properly ordered society.

The *Vedas* declare that humans have the responsibility to function in balance with nature.[4]

To do that, individuals, families, and countries should not become too selfish.

The natural social structure established by universal *dharma* divides society according to the three levels of selfishness.

The *Vedas* dictate that in natural societies, unselfish people must have the most influence. And the others should yield to their guidance, which is to lead society away from selfish comforts and toward service.[5]

When society is not structured in that way, the *Vedas* declare that it is in chaos and against the natural order.[6]

Following the *dharma* that is established for humanity is the basic requirement for all. That *dharma* includes, among other things, truthfulness, minimal violence, minimal exploitation of nature, and the practical acknowledgment of indebtedness.

Such indebtedness is to nature, the governing source, family and lineage, and other beings including animals, plants, trees, insects, etc.

Individuals and societies that follow this *dharma* ultimately connect with the source.

Regardless of theistic conviction, one must follow the *dharma* of nature or the universal order.

Universal *Dharma*

According to universal *dharma*, humans and others are tasked with maintaining order, not destroying it.[7] When humans destroy nature, they are not only violating the physical order of nature by causing an imbalance in ecosystems, but they are also breaking the laws established by the governing source of nature.

Humans should acknowledge their limitations. It is not *dharmic* for humans to neglect maintenance, which means humans must live sustainably with what they are given.

Becoming too absorbed in creative and destructive projects distracts from maintenance.

Naturally, humans are somewhat creative and destructive; however, those who try to excessively expand those tendencies, especially destruction, work against the balance that universal *dharma* establishes.

---

[1] *Manu-smṛti* 8.41: Some *dharmic* levels are *jāti-dharma* (community), *jāna-pada-dharma* (region or country), *śrenī dharma* (guild, association or corporation), and *kula-dharma* (family or dynasty).

[2] *Bhāgavata Purāṇa* 6.1.40: *veda-praṇihito dharmaḥ*, "The *Vedas* prescribe the highest *dharma*. And the *Vedas* are directly Nārāyaṇa, who is the

source of the physical universe."

[3] *Bhāgavata Purāṇa* 6.1.40: "That which is prescribed in the *Vedas* constitutes *dharma*. The opposite of that is *adharma* (against *dharma*)."

Bhāgavata Purāṇa 6.3.19: *dharmaṁ tu sākṣād bhagavat-praṇītam*, "*Dharma* is enacted directly by the monotheistic source of the universe."

Manu-smṛti 2.6 "The root of *dharma* is in the original *Vedas*, their supplements, the character of those who know the *Vedas*, the conduct of those who follow the *Vedas*, and that which gives self-satisfaction."

[4] *Manu-smṛti* 12.88-90: "Responsible work is based on debts to the higher powers, ancestors, society, and other beings in nature."

[5] *Ṛg-veda* 8.4.19, *Yajur Veda* 34.11, *Atharva Veda* 19.66, and *Bhāgavata Purāṇa* 11.5.2 declare that those devoted to the supreme source are naturally at the head of society. They are the most knowledgeable and least selfish.

If one does not have such qualities, he does not deserve to be in a prominent position. (*Bhāgavata Purāṇa* 7.11.35)

[6] *Bhagavad-gītā,* Chapter 16.

[7] *Dharma* is from the verb root *dhṛ,* which means to maintain. Thus, *dharma* on the highest level is established to maintain the universe. The laws of physics, biology, society, nature, etc. are all contained within universal *dharma*. Those laws and more are for maintaining the universe, including nature.

Following universal *dharma* is not optional. There are consequences for breaking *dharmic* laws.

Time

According to the *Vedas,* time is an aspect of the causal source. An understanding of time sheds light on transcendent reality and life's purpose.

Time is an integral part of nature and *dharma.*

Cyclic Time

The *Vedas* explain that time is cyclic.

Cycles exist everywhere in the universe.

The passage of days and nights forms a basic cycle. Seasons and years are larger cycles.[1]

Many thousands of years comprise a *yuga*. One *yuga* is part of a four-*yuga* cycle of four million years. Then 71 of the four-*yuga* cycles make up an eon of 300 million years, continuing to even larger cycles.[2]

There is also a cycle of life. The life force assumes a form or body. That form is born, grows, declines, and dies. But that is not the end. The life force continues the cycle into another body.[3]

The *Vedas* recommend that the life force, or the person, should take part in the cycle of universal maintenance. That maintenance cycle determines the direction of the life force as it cycles through birth, death, and rebirth.[4]

Time as a Force

The *Vedas* say time is both a measurement and a force.

By his glance, the supreme source injects time into the physical repository, which then causes it to unfold, sustain, and withdraw.[5] In that way, time is the force of creation, maintenance, and destruction.

Destruction indicates that time is also the force of entropy or decline followed by dissolution.[6]

---

[1] In popular depictions of the Christian New Year, the old year is an old man, and the new year is a baby. (Wikipedia, "Baby New Year")

This is a rebirth concept. Understanding cycles in time makes it easier to understand how the life force passes from one body to another as part of the cycle of birth, death, and rebirth (*saṃsāra*).

[2] Wikipedia, "Hindu units of time."

[3] *Bhāgavata Purāṇa* 6.17.18: "In the cycle of *saṃsāra,* the life force, veiled by ignorance, wanders through various bodies, sometimes enjoying and sometimes suffering."

Bhagavad-gītā 2.13: "As the life force passes through the forms of childhood, youth, and old age, it also passes into another form after death. The wise are undisturbed by this."

[4] Entwined in the cycle of life and the seasons is a food-sacrifice cycle which connects humans to the maintenance of the universe. (*Bhagavad-gītā* 3.16)

The food-sacrifice cycle begins with humans working for food and other necessities. The controllers of nature supply those necessities. Humans are then obligated to offer sacrifice to those controllers as part of the cycle.

The cycle repeats as humans work, reap the results of work, and use those results in sacrifice again and again.

Doing so is the human responsibility in the maintenance of the universe. It is their small role.

That food-sacrifice cycle is basically the *karma-yajña-saṃsāra* cycle, which is the cycle of life. The cycle of life—birth, death, and rebirth (*saṃsāra*)—is linked to *karma* (human labor) and *yajña-cakra* (the food-sacrifice cycle).

Those who participate in the food-sacrifice cycle are benefitted in the cycle of life (*saṃsāra*). Those who do not, return (cycle) to undesirable forms.

This is the concept of *yajña*, which is explained in a topic below.

[5] *Bhāgavata Purāṇa* 12.3.26: "Nature is set in motion by the force of time."

Bhāgavata Purāṇa 3.26.17-18: "The supreme source (Bhagavān) sets the universe in motion as the time factor. He exists externally as time and internally (pervading the universe) as *paramātmā*."

Paramātmā means the aspect of the source that is omniscient, omnipresent, and the supreme enjoyer.

[6] *Bhagavad-gītā* 11.32.

Rejection of The Source

Many people speculate that a governing source probably does not exist. They declare that there is a lack of proof.

However, their hypothesis is dangerous. It gives society a license for exploitation of nature.

On the one hand, humanists have discarded belief in a governing source, and they have constructed a workable doctrine wherein humans do not treat each other too abusively. That is observable.

On the other hand, no such experience exists regarding human relations with nature.

In other words, without a governing source, humans may succeed in embracing humanistic morals. However, there is no evidence that, without reverence for a governing source, humanity will ever stop over-exploiting nature.

The concept of a source is intuitive among most humans.[1] And it can be easily argued that, historically, societies with reverence for a governing force and its principles have functioned in balance with nature, far more so than those that have discarded them.[2]

External Control

An essential question is whether greed can be sufficiently controlled by human will, or is an external governing power required. The *Vedas* declare that human will is insufficient.[3]

Nature has examples of the difference:

Ecosystems maintain equilibrium. They impose a structure that is external to species, who, left uncontrolled, would try to consume too much in an ecosystem. A species that is not externally governed can wreak havoc on the equilibrium and even destroy the ecosystem.

Therefore, to declare that human greed can be governed by anything other than reverence for an external controller is unnatural and unproven.

Leaders and educated people should not ignore the governing source and his laws. Doing so is a dangerous experiment.[4] The risk is too high for any reasonable, responsible person.

In a free society, people may certainly place their faith in a random source. However, when society is being led to ruin, such irresponsibility must be checked.

Many people declare that they will wait for proof of a governing source.[5]

However, such waiting does not cause them to pause before violating universal *dharma*. The *Vedas* state that one who overconsumes nature's resources is a thief who steals from nature.[6] By rejecting that *dharma,* one becomes an outlaw.

Thus, waiting is not waiting. It is acting, acting in violation of nature's law.

---

[1] "The Global Religious Landscape," Pew Research Center, PewResearch.org: "Worldwide, more than eight-in-ten people identify with a religious group."

"Intuitive Thinking May Influence Belief in God," American Psychological Association, apa.org.

[2] The statement here is "Societies with reverence for a governing force and its principles have functioned in balance with nature."

This includes societies centered on a deterministic universe but without a clear definition of the source. Polytheists orient society around many controllers. Nontheists, like Buddhists and Confucianists, also accept a deterministic or governed universe.

They all reject atheism, which implies random or probabilistic causes in the universe. The point is that one may argue about the source or its methods of governance. However, one should not accept the destructive anarchy of atheism.

In the book, *The Environmental Solution—Overconsumption is Misplaced Enjoyment* by Michael Wright, there is a comparative analysis of various cultures. The book demonstrates that cultures that follow the laws of a governing force are more in balance with nature than modern atheistic culture.

3 *Bhagavad-gītā* 2.59: "One may be able to restrain oneself from desires, but until one experiences something higher, the taste (inclination) will remain."

4 "Experiment," meaning a society that is governed by unproven, predominantly atheistic systems like modern democracy, humanism, and science.

Although some adherents of those systems declare that they include an acceptance of a causal or governing source, the systemic disregard for *dharma* that is based on a supreme source nullifies their faith.

5 As agnostics.

6 *Bhagavad-gītā* 3.12 and *Bhāgavata Purāṇa* 7.14.8.

Comfort and Satisfaction

In modern society, due to the predominance of selfishness and extreme selfishness, people are generally dissatisfied.[1]

They seek satisfaction from comforts achieved through overconsumption. However, those comforts do not provide true satisfaction. Instead, they mask dissatisfaction.

That mask is the illusion created by *māyā*. Many are intent on that dynamic. That is despite old age, disease, and death that signal it is not a practical direction. Another signal is the one given by environmental destruction caused by accumulating too many comforts.

Social Satisfaction

Achieving satisfaction is not just an individual prerogative. It must be societal.

A society must be strongly influenced by a class or

group of people who are illuminated. They may not be purely so, but they should be examples of service and self-satisfaction.[2]

---

[1] Selfishness and extreme selfishness are dimness and darkness, *rajas* and *tamas* in Sanskrit.

[2] *Bhāgavata Purāṇa* 11.17.16: Self-satisfaction (*tuṣṭa-ātmā*) is one of the natural characteristics of an illuminated person.

Modern Culture

For info on Zoom meetings, email info@cvc.guru.

Introduction

The previous section discussed the self as the life force. When one identifies as life force, the objective is not to fully divorce oneself from the body. Instead, one uses it to serve nature and the governing source.

Modern society is largely based on secularism and socio-economic ideologies. As a result, systems are created that are rooted in physical reality. Those systems make the perception of transcendent reality optional. They generally work against reverence for the source and his laws.

In Vedic culture, the reverse is true. Social and governmental systems are based on universal *dharma,* which is in turn rooted in transcendent reality.

This section analyzes various institutions in modern culture to determine whether they establish balance with nature.

The essential question is whether they encourage overconsumption, control it, or do nothing about it.

Science

Science and Reality

Empirical science creates a perception of reality based on observing the physical world, meaning the universe and nature.

Solely identifying with physical reality, which is akin to animal reality, is dangerous. It is generally an impetus to violence, war, and overexploitation of natural resources.[1]

Currently, identification with physical or empirical reality limits the perception of highly educated and influential people of the world.[2]

They do not represent most people on earth,[3] but they have the most influence and their voice is the loudest.

Science is Human-centric

Human-centrism is harmful to nature because its view is that the center of nature is humanity. In that view, the components of nature are objects that may be possessed, exploited, or destroyed depending on human desire.[4]

However, human-centrism is an opinion or a belief like a religious belief. Some say that modern science, which is an outgrowth of the European renaissance is influenced by the human-centrism of Christianity.[5] That is also the position of Vedic cultural ecology.

Vedic cultural ecology asserts that human-centrism defines a perception of reality that is contrary to nature's reality.[6]

Thus, human-centrism is unnatural and selfish.[7]

Science and *Dharma*

Science defines a type of *dharma*, which is a set of rules or laws. The laws that science assembles are those of the origin, function, and demise of the physical universe.

However, despite science's powerful and ubiquitous influence in society, its *dharma* does not address society's greed. Due to that omission, scientists are complicit in the crime of overexploitation of nature.[8]

For that reason, Vedic cultural ecology views modern science as immoral. In contrast, most scientists would declare that science is amoral, meaning without any morals.[9]

In other words, they think that science is above the need for morals.

Despite that opinion, the *Vedas* assert that scientists must be unselfish examples in society, and they must uphold *dharma* that establishes unselfishness and the control of greed. If they do not do so, they cannot be taken seriously as scientists.

Moreover, scientists must guide governments so that they properly enforce universal *dharma* in society.

Those features are included in the Vedic definition of *dharmic* science.[10]

---

[1] *Bhagavad-gītā* 16.8-9: "People who identify only with the physical and ignore its source lose their ability to discern reality. Their lack of discrimination leads them to acts of violence, which ultimately threaten to destroy the world."

"Emergence of War in Plato's Republic," by Olivia Garard, thestrategybridge.org: "Feeding one's citizens is a basic need. To seek beyond, misprioritizes needs because it introduces and integrates other desires among the basic requirements.

"For 'if they let themselves go to the unlimited acquisition of money, overstepping the boundary of the necessary' then, Socrates posits, the other city will seek more land, too. War results from this interactive acquisitiveness." (Plato, The Republic, 373d)

[2] "When Science Meets Power," by Geoff Mulgan: "Politicians often assume that 'following the science' will help make choices more straightforward."

"Scientists and Belief," Pew Research Center, pewresearch.org: A survey of scientists finds that half believe in some form of deity or higher power. By contrast, 95% of Americans believe in a deity or higher power.

[3] Wikipedia, "Demographics of atheism:" "According to the Pew Research Center, 16% of the world's population is not affiliated with a religion, while 84% are affiliated.

"Furthermore, many of the unaffiliated, which include atheists and agnostics, still have various religious beliefs and practices."

[4] Wikipedia, "Anthropocentrism."

[5] The Bible, Genesis 1.26-28.

[6] The *Vedas* give nature's perspective:

Bhāgavata Purāṇa 12.3.1: "Seeing the controllers of men busy trying to conquer her, the earth laughed. She said: 'Look at how these controllers, who are playthings in the hands of death, desire to conquer me.'"

Bhāgavata Purāṇa 7.15 "Upon seeing a person who is killing animals, the animals are afraid: 'This merciless exploiter of animals, ignorant of the purpose of life, is going to kill us.'"

But humans are mostly interested in the human-nature relationship from the human point of view. That is because they do not see nature as sentient, and thus they believe it has no perspective.

There may be two ways to see the world from nature's view.

The first is to understand that nature is indeed sentient.

The second way is to see nature as neutral toward all beings. If nature is neutral, then it is not logical or scientific to declare that humans are special.

[7] "What is anthropocentrism? (A definition)," *The Ecological Citizen: Confronting human supremacy*, ecologicalcitizen.net: "Anthropocentrism

(human-centrism) suffers from crippling limitations, including the power of selfish interests, a narrow frame of reference and view of relevant non-human stakeholders, a short-term time scale, our (human) ignorance and our fallibility."

[8] In the U.S., school shootings have become a problem. Recently, authorities have begun prosecuting the parents of child murderers. They are charged with second-degree murder or manslaughter.

Vedic cultural ecology argues that scientists are similarly complicit. They cannot supply technologies that are effectively weapons of mass destruction of nature and remain free from culpability.

[9] "The Impact of Moral Values on the Promotion of Science," by Hassan Zohoor, *National Library of Medicine*, ncbi.nlm.nih.gov.

[10] *Īśa Upaniṣad* 9: "Those who cultivate ignorance enter the regions of darkness. But worse are the leaders of society who cultivate only physical knowledge."

Bhagavad-gītā 3.21: "The common people follow the leaders of society. Whatever standards they set the world pursues."

Medicine

The scientific revolution introduced advances in medicine alongside other branches of science.[1] And the lives of most are deeply affected by medical developments.

Thus, medicine is often considered to be one of the pillars of modern science and technology.

Its advances have led to better health and longevity. And many consider them to be the ultimate proof of science's power and importance.

However, medical science causes selfish human-centrism.[2]

Too much emphasis on health increases identification with the body and mind.

In contrast, those who identify with the life force are the best at tolerating nature, including its diseases.[3]

Most people in Vedic society consider identification with the life force and transcendent reality to be ideal.

As a result, doctors are indeed respected in Vedic society; however, medical science is not deified like it is in modern society.

It is culturally unhealthy for humans to think they should conquer nature and destroy all diseases, which are part of nature. The Vedic priority is to live in balance with nature and tolerate its harshness and diseases as much as possible.

---

[1] The Scientific Revolution, developing between the mid-1500s and mid-1600s, corresponds roughly with European Renaissance, with John Locke and Issac Newton playing important roles.

[2] One can do a search on "medical science and human centric" and find many responses. Of course, medicine is for humans.

However, Vedic cultural ecology is making a different point. Medical science or the health industry increases humanity's identification with physical reality. That is because physical reality for the individual is centered on the physical body.

[3] *Bhagavad-gītā* 2.14-15: "You must tolerate the temporary physical events that come and go. Doing so will liberate you from suffering."

Bhāgavata Purana 5.14.27: "Physical reality is characterized by happiness, distress, attachment, hate, fear, false prestige, illusion, madness, lamentation, bewilderment, greed, envy, enmity, insult, hunger, thirst, tribulation, disease, birth, old age, and death. These all combine to produce misery."

Bhāgavata Purāṇa 5.14.39: "Considering the miserable condition of physical reality, one should seek the company of self-controlled sages who are self-satisfied and friendly toward all beings. They can award liberation."

Rights

This topic is especially relevant to the following chapters on democracy, humanism, socialism, and capitalism.

The Right to Work

Humans have the right to work for existence, knowledge, and enjoyment. However, the *Vedas* explain that nature does not give humans the right to achieve them.[1]

In other words, nature may either thwart or facilitate the human desire for enjoyment, knowledge, and life itself.[2]

Moreover, the *Vedas* explain that living beings do not have the right to live free from exploitation. Almost all beings are food for others.[3] And they exploit and kill each other for various reasons.[4]

Among humans, only *dharma* prevents them from violence. Non-violence is not automatically a part of human nature, which mirrors animal nature.

Obsession With Rights

Humans often strive to implement certain rights.

In 1948, the United Nations adopted a "Universal Declaration of Human Rights."

Those rights include freedom, security, property ownership, democratic participation, leisure time, sufficient pay for work, and free education.

Such rights have established some *dharmic* guidelines and controls for governments. But they go too far. Humans think their man-made *dharmas* supersede universal *dharma*.

In contrast, the *Vedas* declare that nature's *dharma* governs humans.[5] People in modern society spend too much time trying to establish utopia on earth. They ignore nature, which does not support such utopia.[6]

The problem is not understanding that humans do not control nature.[7] They cannot force nature to award them rights. It is better to spend less time trying to assert rights and more time learning how to live as nature's subject.

---

[1] *Bhagavad-gītā* 2.47, "You have the right (*adhikāra*) to work for any result. But do not think that you are entitled to achieve it. You have no control over whether you may achieve it."

[2] Here, "thwart life itself" is a conventional expression for death. Nature may destroy one's body, but the life force is not destroyed.

[3] Even predators at the top of the food chain become food for various parasites.

[4] *Bhāgavata Purāṇa* 1.13.47, "The weak take advantage of the strong. One being is food for another."

[5] *Gautama-dharma-sūtras* 11.20-21.

[6] Suffering, disease, and death are natural.

The *Vedas* state that governments have a responsibility to relieve human suffering; however, they are not above nature. Citizens should not be under the impression that the government is responsible for the cause of their suffering and enjoyment.

Having that impression makes them ignorant. And they typically feel entitled to more than nature allows.

[7] *Bhagavad-gītā* 7.14, "Nature is extremely difficult to overcome. But by connecting to its source, one may surmount the illusion (*māyā*) of physical reality."

Democracy

Democracy and Reality

Democracy generates a social reality, fashioned by the collective action of the public. It is a type of physical reality. Experts state that talk, votes, and markets create that social reality.[1] And the joint action of citizens establishes the law.[2]

The reality created by democracy is heavily dependent on business and government, which, in large part, are formed by the desires of the public.

For many in democracies, the government comprises an inordinate part of their perception of reality. They think their opinions and votes will influence their fate. They become intoxicated with their control.[3]

Some have so much faith in and reliance on democracy that it disturbs their reality when they see it working imperfectly.[4]

A perception purely based on democracy leads people to believe the state or nation is the most important part of reality. However, that perception ignores the fact that nature and its governing source are above the state.

Democracy and *Dharma*

A democratic system relies on a consensus about what principles (*dharma*) lead to the enactment of laws. However, modern global culture, in which democracies thrive, is generally based on selfish human-centrism.

People who are addicted to overconsumption are unlikely to stop themselves in a democracy. Some may try to pass laws against it, but the majority are unlikely to agree. Or if they do, the laws will not be effective.[5]

From the Vedic perspective, democracy is neither good nor bad, but the majority who rule themselves must follow nature's *dharma*. Without it, they are the blind leading the blind.[6]

---

[1] "The Epistemology of Democracy," by Elizabeth Anderson, *Episteme* 3.1–2.

[2] "The law"—*dharma*.

[3] "Why America's Intoxication with Democracy Can Be Extremely Dan-

gerous," by John Ashcroft and David Lane, Hughs News, hughsnews.com.

This discussion does not precisely correspond with Vedic cultural ecology's position on the intoxication of democracy; however, it is roughly similar.

[4] Wikipedia, "Democracy:" *Creative Democracy: The Task Before Us* by John Dewey. "Dewey argues that democracy is a way of life and an experience built on faith in human nature, faith in human beings, and faith in working with others."

[5] Wikipedia, "Prohibition in the United States."

[6] *Kaṭha Upaniṣad* 1.2.5: "People go to leaders for guidance, even though those leaders are ignorant. As a result, they both suffer like the blind leading the blind."

Humanism

Humanism is almost synonymous with human-centrism, which is a type of physical reality.

Many say modern humanism comes from Christianity.[1] It rejects the idea of God but tries to keep most of the Christian morals concerning human relations.

Humanist Reality

Modern humanism began in the 19th century. And since then, it has become ubiquitous.

In general, modern humanists believe in democracy. They try to be moral toward other humans. They have faith in human ability.[2]

Although many state their desire to preserve nature, their perception of reality prevents them from fully doing so. When humanists and leftists are forced to choose between nature and humans, they typically choose the latter.[3]

Over the last century, humanist perceptions have increasingly pervaded modern global culture.

In contrast, the *Vedas* support humans, but not at the expense of the life force, nature, or *dharma* established by the governing source.[4]

Humanist *Dharma*

Although humanists decry the exploitation of others, they ironically try to colonize the world with their ideals and standards.[5]

That means anyone caught violating so-called human rights is condemned.[6]

To establish humanist *dharma,* they must imagine how humans would behave in a utopia or ideal society. Based on that speculation they derive goals, values, morals, and laws.

Not all their principles contradict universal *dharma,* but many do.

Humanists believe they can control greed through education, laws, and improved socio-economic systems. Thus, they establish a *dharma* accordingly.

Yet, their *dharma* has not controlled the human greed that causes overexploitation and destruction of nature. Instead, they offer only empty promises for a brighter future.

---

[1] Wikipedia, "Humanism:" "Anthropology professor Talal Asad argues humanism is a project of modernity and a secularized continuation of Western Christian theology."

Also, "According to Didier Fassin, humanism originated in the Christian tradition."

[2] Wikipedia, "Humanism."

[3] "Firms should not penalise meat eaters, warns TUC," *BBC News*, 16 July 2018.

This article is about a memo issued by WeWork, a company of 6,000. It said, "Moving forward, we will not serve or pay for meat at WeWork events..."

That was due to the company's concern for the environment.

But the response from the TUC (Trades Union Congress) in England was that firms should not penalize workers.

The union, which is indeed mandated to protect workers, chose the desires of the workers over concerns for the environment.

Although this is only one example, it illustrates the dichotomy between the discipline required to minimize consumption and the natural, yet animalistic, desire to live at the expense of nature.

[4] The *Mahābhārata, Vana Parva,* contains a story that reveals some attitudes and behaviors in Vedic culture:

A pigeon, who was chased by an eagle, took shelter of King Śibi. But the eagle declared that a pigeon is his natural food, and the king had no right to protect it.

As a result, the king offered to give his own flesh, which the eagle accepted.

The story is longer; however, it teaches a few lessons:

It instructs that the greatest among us is charitable to a fault. It teaches that even a king must respect the order of nature. And it shows that the government is responsible for protecting nature.

Although this story is little known outside Vedic culture, nearly everyone in the culture, hundreds of millions of people over thousands of years have heard and learned from this tale.

[5] "Exploring Perspectives of Humanism," by Muhammad Arif Khan, *The Dialogue* 14.1.

Khan explains that humanism, despite its liberal proclamations, carries a tendency toward colonization. By that, he means establishing doctrinal control. He states that its propagation of values, norms, and traditions is all-pervasive in modern global culture.

[6] Wikipedia, "Human rights inflation."

Socialism

Socialism and Reality

Socialism defines a type of social reality.[1]

Most socialists see two such realities: an oppressive one that describes the status quo and a socialist one that introduces an ideal social reality, which is a kind of utopia.[2]

Both capitalists and socialists or liberals strive to improve their material condition because they think reality is physical and social.

Some mix their perceptions by acknowledging a higher power, perhaps nature or a god.[3] Nevertheless, even those people tend to be anchored in the physical perception of reality.

Socialism and *Dharma*

Socialists claim the problems with nature are the fault of the capitalists.[4]

The socialist idea is not so much to reduce consumption, but rather to control and direct it through laws and science. They still want to exploit nature, but through means that will not be as destructive and polluting.

Also, they point out that the world's richest are the main exploiters, which is true.[5] Their idea is to redistribute wealth and reduce poverty as part of a joint solution that helps both nature and humans.

However, their idea of linking poverty with excess exploitation of nature is not proven. It serves socialist ideology, not nature.[6] Poor people are not any less greedy than rich ones. The poor simply do not have the means to be as greedy as the rich.

There are examples of countries moving from poverty to wealth.[7] They do not want to be controlled by European socialist or liberal policies on the environment.

In sum, socialist *dharma* is deficient.

---

[1] Vedic cultural ecology puts liberals and those influenced by socialism under one heading due to their similar perceptions of social reality.

[2] "Some Comments about Marx's Epistemology" by Prabhat Patnaik, *Monthly Review Online*, mronline.org.

Oppression of the proletariat (workers).

Wikipedia, "Socialism."

---

[3] Wikipedia, "Religious socialism."

[4] "Capitalism and the Environment," by Geoffrey Jones, *Evolutions of Capitalism*: "Capitalism drove the environmental decimation of the planet."

[5] *How The Rich Are Destroying the Earth* by Hervé Kempf, amazon.com.

The richest include North Americans, Western Europeans, Japanese, Australians, South Koreans, etc. ("Median income or consumption," Our World in Data, ourworldindata.org)

[6] "Merging the Poverty and Environment Agendas," by Paul Delia, *International Institute for Sustainable Development*.

[7] India and China are two examples.

"China Abandons Paris Agreement, Making U.S. Efforts Painful and Pointless," by Diana Furchtgott-Roth, The Heritage Foundation, heritage.org.

"India's power minister accuses west of hypocrisy over energy transition," by Benjamin Parkin, *Financial Times*, ft.com.

Capitalism

Capitalist Reality

Like democracy and socialism, capitalism establishes a physical or social reality. Prof. Elizabeth Anderson explains that in a democracy, social reality is formed through talk, votes, and markets.[1]

That conforms with the view of capitalists. In general, they perceive an economic or social reality within the context of market forces.[2]

Some capitalists are more philosophical than others. They believe capitalism is not just about money but any form of individual or social value. Such capitalists may be more philanthropic than others. In that way, they believe that stabilizing the lives of other humans brings material worth to society.[3]

In any case, the goal of pure capitalism is physical or material worth. In contrast, the *Vedas* establish detachment from physical worth.[4]

Capitalist *Dharma*

Capitalist *dharma* is based on the various practical functions used to increase material wealth.

In Vedic culture, there are capitalists and socialists; however, they must both conform to universal *dharma*.[5] Such *dharma* restricts the greed of both capitalists and the proletariat.

---

[1] "The Epistemology of Democracy," by Elizabeth Anderson, *Episteme* 3.1–2.

[2] Wikipedia, "Capitalism."

[3] "Variants of Epistemic Capitalism: Knowledge Production and the Accumulation of Worth in Commercial Biotechnology and the Academic Life Sciences," by Maximilian Fochler, *Science, Technology, & Human Values* 41.5

[4] *Amṛta-bindu Upaniṣad* 2: "The mind is the cause of both bondage and liberation. When it is attached to physical reality, it is the cause of bondage and suffering, and when it is detached, it is the cause of liberation."

[5] *Bhagavad-gītā* 18.28, 18.46-47, and 2.33.
Bhāgavata Purāṇa 7.15.14 and 11.18.46.

Advancement

A predominant behavior in modern society is the continual quest for development. Year by year, societies seek to advance their physical knowledge, technologies, economies, quality of life, etc.[1]

Such behavior is selfish.[2]

It is natural for humans to be creative; however, when society is ruled by development or so-called advancement, it is out of balance.[3]

Instead, the predominance of maintenance is required in a healthy society.[4]

Advancement

To achieve advancement, humans exploit other humans and nature.[5] The pursuit of advancement is deeply rooted in physical reality.

Most people assume technology is the measure of an advanced society. That assumption exists throughout science and modern global culture. But it is based on devotion to comforts.

Modern scholars often judge previous cultures primarily based on their technical advances.[6]

Advancement is also related to quality of life. That typically means how much comfort is available through advancements.[7]

In sum, the constant pursuit of advancement disrupts the proper maintenance of society and nature. Moreover, acquiring comforts through advancement does not provide true satisfaction.

Economic Expansion

The following describes in brief how consumption grew from a benign stage to a malignant level that now harms the whole earth.

A thousand years ago, Europeans began to trade beyond Europe.[8] That slowly led to improved navies and armies. In the 1500s, they began to establish colonies.[9]

In the 1800s, due to scientific discoveries, their industries grew stronger,[10] which in turn strengthened their armies and businesses.[11]

Christian impact on Western society lessened.[12] Over the centuries, the controls on greed gradually weakened.[13]

Democracy became popular.[14]

A new focus on the rights of the people emerged. Socialism appeared in response to the industries that were exploiting workers.

In the 1920s, socialism and capitalism jointly fostered a consumer economy. That was caused by the greed of capitalists who took advantage of the newly unleashed desires of the better-paid masses.

And mass consumption led to the overuse of natural resources.

In the late 20th century, such abuse of nature caused serious pollution and a lack of sustainable resources.

The result is that in the 21st century, the consumption rate cannot sustainably continue.[15]

Since the socialists and capitalists have caused the problems, they cannot solve them. They falsely claim to have that ability. But they cannot change their core values.

---

[1] Wikipedia, "Economic development," "Technological change," "History of science," "Progress," "Invention," "Developing country," "Development theory," and "Human Development Index."

[2] *Bhāgavata Purāṇa* 11.18.26: "One should not accept the perishable phys-

ical reality as the ultimate. Instead, one should be detached from physical objects along with progress or development in the physical world."

In current thinking, development is usually not considered to be selfish provided people act responsibly. However, the *Vedas* explain that such a sense of responsibility should be discarded because it contradicts the higher *dharma*. The following are examples of modern thought on selfishness:

"Does American Society Encourage Selfishness?" by Thomas Henricks, Psychology Today, phychologytoday.com: "American social institutions (like economics, family relations, and healthcare) support a private, even selfish vision of life."

"Global evidence on the selfish rich inequality hypothesis," by Almas, Cappelen, et al, National Library of Medicine, ncbi.nlm.nih.gov: "We provide global evidence showing that where the fortunes of the rich are perceived to be the result of selfish behavior, inequality is viewed as unfair, and there is stronger support for income redistribution. However, we also observe that belief in selfish rich inequality is highly polarized in many countries and thus a source of political disagreement that might be detrimental to economic development."

[3] Although seemingly delightful, creation, birth, and creativity are on the second level of selfishness, dimness. Those functions may indeed be encouraged but controlled or guided.

[4] *Bhāgavata Purāṇa* 11.4.5, 10.24.22, 2.5.18, and 11.22.12: Maintenance is the most *dharmic*. Creation is second (*rajas*) and destruction is the lowest (*tamas*).

Bhagavad-gītā 14.6: "The mode that sustains is purer than the others. It is enlightening and liberating."

Bhagavad-gītā 14.16-18: Actions that sustain are the best. They are faultless, pure, and congruent with the universal *dharma*.

[5] *Bhāgavata Purāṇa* 1.13.47: "The strong prey on the weak. One being is

food for another."

[6] "The Place of Technology in Civilization" by Fred Hoyle, *Engineering and Science* 16.5 (1953): "...technology controls civilization, and the details and variations in social organizations are relatively unimportant, except where the social organization in some degree affects technology itself."

[7] Wikipedia, "Quality of life."

[8] Wikipedia, "Commercial revolution."

[9] Wikipedia, "Analysis of European colonialism and colonization."

[10] Wikipedia, "Scientific revolution."

[11] Wikipedia, "Industrial revolution."

[12] Wikipedia, "Decline of Christianity in the Western world."

[13] "Greed Is Good: A 300-Year History of a Dangerous Idea," by John Paul Rollert, *The Atlantic*, 2014.

[14] Wikipedia, "Democratic transition."

[15] "The Pandemic of Consumerism," *UN Chronicle*, by Jorge Majfud: "Trying to reduce environmental pollution without reducing consumerism is like combatting drug trafficking without reducing the drug addiction."

Religion

Jesus said, "Do not store up for yourselves treasures on earth, where moths and vermin destroy, and where thieves break in and steal. But store up for yourselves treasures in heaven, where moths and vermin do not destroy, and where thieves do not break in and steal."[1]

Most religions consider greed a vice. And for many hundreds of years, they were successful in guiding their followers away from too much greed.

Of course, it always exists in humans and cannot be easily controlled.

However, over the last few hundred years, the Christian churches, especially in the West, have gradually lost their ability to control the greed of their followers.[2]

Colonialism and overexploitation of natural resources began in Europe's Christian countries. Today, people of all religions are infected to a greater or lesser degree. They want a share of the treasures being stolen from nature.

And many, not just Christians, believe it is their God-given right to overexploit nature.[3]

Greed is a vice that should be controlled, especially when it causes the degradation of nature. All of society's religions, ideologies, philosophies, and sciences have a responsibility to control greed and overconsumption.

---

[1] *The Bible,* Matthew 6:19-20.

[2] Wikipedia, "Decline of Christianity in the Western world," "Christianity and colonialism."

Also, "Pope Speaks Out Against Greed, Apologizes for Colonialism," by *Voice of America News,* voanews.com.

[3] Prosperity theology is the Christian belief that God wants his followers to be wealthy in this life. It took hold in the 20th century and has continued to grow in this century.

See, "Three Out of Four Christians Now Believe in the Prosperity Gospel," Aug 23, 2023, relevantmagazine.com.

Wikipedia, "Prosperity theology."

Entertainment

Arts and Politics

Many artists, actors, and musicians want to improve human relations with nature. They often believe the public needs to be aware of the issues.[1]

Liberals believe education is vital in solving pollution and climate change. But simply teaching that problems exist is not enough.

Education in principles is required.[2] Those principles must include training in how to control the desire to overconsume.

Encouraging awareness and changing attitudes barely approach a solution. There must be modification of behavior on a societal level.

Cinema, Music, Sports, and Games

In traditional societies, theater, music, and sports were local events. Until the advent of mass communications, they were presented on a small scale.

According to Vedic cultural ecology, an institution

or industry that influences the public should agree to one of two paths: It should either guide the public responsibly or it should allow itself to be controlled.

That means it should directly or indirectly influence its followers to consume less and live in balance with nature.

However, from top to bottom, that is not happening in the entertainment industry. Nearly all the entertainment topics and themes encourage identification with physical reality.

The lifestyles of the highly paid entertainers reinforce examples of overconsumption.

The industries themselves overexploit, waste, and pollute nature.[3]

Most importantly, the people who are entertained—the fans—must work hard exploiting the earth to earn money for their entertainment pursuits.[4]

In Vedic culture, the ancient seers warned against entertainment that was based solely on physical reality.

Administrators restricted the payment of entertainers and controlled their performances so they would not impact the smooth functioning of village life.[5]

Buddha, Śaṅkara, and others outlawed it for their

followers.[6]

However, many Vedic traditions encourage music and entertainment when they relate to transcendent reality.[7]

In any case, simple, local drama, storytelling, singing, and sports do not devastate the environment.

---

[1] "What is Environment Awareness?" Vinciworks, vinciworks.com.

"'Washed Up' Art Exhibition Raises Awareness of Plastic Pollution," Wilson College, Wilson.edu.

[2] Here, *principles* refers to *dharma*.

[3] "Film and TV's Carbon Footprint Is Too Big to Ignore," by Sarah Sax, Time, time.com.

[4] See, for example, Wikipedia, "Green gross domestic product."

The concept that a higher GDP (gross domestic product) is directly related to higher environmental footprint and loss of biocapacity is somewhat disputed.

However, Vedic cultural ecology's research indicates that the two are related.

[5] *Kauṭilya Artha-śāstra* 2.1: "Entertainers should not receive excess payment or disturb the village work."

[6] *Ghitassara Sutta* Anguttara Nikaya 5.209: the Buddha said: "Monks, there are five dangers of reciting the Dhamma with a musical intonation.

"What five? Oneself gets attached to the sound; others get attached to the sound; householders are annoyed, saying, 'Just as we sing, these followers of the Sakyan sing;' the concentration of those who do not like the sound is destroyed, and later generations copy it."

Caitanya-caritāmṛta Ādi 7.41: "A monk who follows Śaṅkara should study. He should not sing and dance in praise of the monotheistic source, Viṣṇu."

Note that these are injunctions regarding *mantras* and hymns. Ordinary music was completely rejected.

[7] *Bhāgavata Purāṇa* 7.5.23-24 and 7.7.30-31: "One should hear and chant about the *līlās* and names of Viṣṇu."

Bhāgavata Purāṇa 2.6.38: "Śrī Brahmā said, 'Those who are the most exalted in the universe sing of Viṣṇu's *līlās*.'"

A Broken Machine

When repairing a broken machine, a good technician determines the root cause of failure.

Moreover, when a working model is available, one may fix the broken machine by comparing its components with ones that function properly.

Such is the methodology of Vedic cultural ecology. It compares broken modern culture with properly functioning Vedic culture.

In general, that is the practice of modern medicine. In simple terms, blood components in a sick person are compared with those same components from healthy people.[1]

---

[1] Bain, BJ; Bates, I; Laffan, MA (2017). *Dacie and Lewis Practical Haematology*: The complete blood count is interpreted by comparing the output to reference ranges, which represent the results found in 95% of apparently healthy people.

Vedic Culture

For info on Zoom meetings, email info@cvc.guru.

The Supreme Source

In Vedic culture, two prominent groups identify with transcendent reality and seek to connect with its source: Vedic monotheists and Vedic pantheists.

Vedic monotheists are about 70% of the billion people in Vedic culture.[1] They embrace family and monastic life that practices detachment from the physical perception of reality. Most importantly, they are attached to the supreme source, Viṣṇu.

Traditional Vedic pantheists place more stress on monasticism or total disconnection from family life. They are highly philosophical and have trouble encouraging a practical balance with physical reality.

Vedic pantheism has difficulty standing alone to create a complete culture like monotheistic culture does. In large part, that is due to their disregard for the *līlās*, which are a natural part of Vedic monotheistic culture.[2]

The *Purāṇas* explain that the *līlās* are eternal and provide a link to the source, Viṣṇu. They do not originate from pantheistic doctrine.

Yet to the chagrin of pantheists, the *līlās* are ubiquitous. And, being the heart of Vedic culture, they cannot be removed.

Since they do not fit naturally in pantheism, its philosophers have an awkward time accommodating them.

Pantheists concoct the idea that they are mystical windows into the pantheistic perception of the source.[3]

They say that people may first enjoy celebrating the *līlās* and then abandon them to go higher. That is like encouraging Romeo to fall in love with Juliet and then renounce her.[4]

The idea of abandoning the *līlās* is disingenuous and manipulative.[5] Requiring people to do so makes the attainment of the pantheistic goal unrealistic for most. It is an elitist doctrine.[6]

In sum, Vedic pantheism is dependent on the monotheistic culture like a benign parasite depends on its host.[7]

Vedic monotheists have the most to offer to modern society, since they have a complete culture with working solutions to overconsumption.

---

[1] *The World's Religions in Figures: An Introduction to International Religious Demography*, by Todd M. Johnson and Brian J. Grim.

[2] In Sanskrit, *līlā* (pronounced leela) means play, sport, pastime, or drama. It refers to the dramas the Supreme enacts when He descends to earth.

See the topic below entitled, "*Līlā* is Enjoyment."

[3] The masses are devoted to the *līlās,* so instead of rejecting them, the pantheists have opportunistically redefined them.

The pantheistic perception begins by analyzing the world as a place of duality—good/bad, etc. That includes the duality of animate consciousness and inanimate matter.

The pantheistic goal is to transcend all duality and achieve oneness.

That goal is achieved through acquiring spiritual knowledge and practicing renunciation—that is, renouncing the world of duality.

[4] The purpose of the *līlās* is to link with Viṣṇu through the love experienced in the *līlās*.

[5] Although the *līlās* are dramas enacted by the source, Viṣṇu, the pantheists classify them with the duality that must be transcended along with physical reality.

The fact that they classify the *līlās* adjacent to physical reality adds insult to injury.

The Vedic pantheists depend on the host monotheistic culture.

And then, like bad guests, they devalue the source, Viṣṇu, along with His *līlās,* which comprise the core of the host culture.

Moreover, their unwillingness to exalt the *līlās* severely weakens the second barrier described below. And that second barrier, which is non-exploitative pleasure, is the most essential of the four.

[6] Pantheism is elitist because it is philosophical, not practical. Its followers attempt to transcend physical reality through acquisition of knowledge and rejection of family life.

In contrast, the masses require a dynamic process that offers enjoyment and love. Celebration of the *līlās* fulfills that requirement.

But the philosophical pantheistic concept is that such love must be "transcended" to merge with Brahman. That idea does not resonate with someone in love. It cannot be heard or tolerated.

In that way, the message is dry and elitist.

[7] Vedic pantheists, like monotheists, work to transcend physical reality or *māyā*. They follow the *Vedas*. In their perverse way, they are respectful of Viṣṇu and the *līlās*.

Thus, the two are "frenemies" or friendly enemies. They do not have physical fights. They may even celebrate and feast together. But they sometimes have heated debates.

Four Barriers, Four Paths

The Solution—Barriers and Paths

Four fundamental aspects of Vedic culture bring society into balance with nature. These aspects create cultural barriers that stand between nature and society's greed.

On the one hand, the aspects prevent overexploitation; on the other hand, they pave the way to satisfaction and balance.

The next four chapters expand on the following brief descriptions of them:

Equality of All Life

The life force in every life form is the same, and it is directly connected to the supreme source. Along with the source, all life is to be revered.[1]

Non-exploitative Pleasure

Līlās, which are at the heart of Vedic culture, offer pleasure that has minimal impact on nature.[2]

Agrarian Economy

Vedic culture is rooted in agrarian life wherein working hard for excess consumption is minimized. Moreover, the Vedic agrarian economy is based on mutual exchanges with Viṣṇu.[3]

Tolerance of Nature

In Vedic culture, people are taught to tolerate the vicissitudes of nature. Such tolerance is part of respecting nature instead of fighting with it.[4]

---

[1] *Bhagavad-gītā* 2.24, 5.18, 7.4, and 18.54.

Taittirīya Upaniṣad 3.5.1: "The life force is Brahman (*brahma*). From that life force, all beings arise."

Taittirīya Upaniṣad 2.1.1-2: "Brahman is existence beyond the designations imposed by physical reality. It is devoid of restriction and unlimited in qualities."

[2] *Bhagavad-gītā* 10.9: "Those whose thoughts dwell in Kṛṣṇa and whose lives are dedicated to service to Kṛṣṇa, perpetually derive great satisfaction and enjoyment from celebrating His *līlās*."

Bhāgavata Purāṇa 1.1.17: "The most exalted sages celebrate the *līlās* of Kṛṣṇa. We are eager to hear about them from you."

[3] *Bhagavad-gītā* 10-16: "Along with the unfolding of the universe, arose *yajña*. Performing *yajña* bestows necessities for living and liberation.

"When offerings are given to the higher powers through *yajña*, those higher powers reciprocate. Through that process humans enjoy life.

"If one tries to enjoy nature's bounty directly without *yajña*, he is stealing from the higher powers.

"All beings must eat; food is produced from rain. And the rains result from performance of *yajña*. *Yajñas* can be performed when one works responsibly and harvests his crops.

"That responsible work is prescribed in the *Vedas*. And the *Vedas* arise from Brahman. Thus, Brahman pervades the *yajñas*.

"One who does not follow the cycle described above lives in vain."

(See the topic "*Yajña*" in the chapter "Non-Exploitative Pleasure.")

[4] *Bhagavad-gītā* 2.14-15: "Happiness and distress arise from sense perception. They are like the passing of summer and winter. One must tolerate them.

"Thus tolerating, being equal in both happiness and distress, one becomes liberated from suffering."

Equality of Life

A Barrier

This is the first of the four barriers that inhibit the overexploitation of nature.

The *Vedas* declare that the life force is equal in all beings. It is an energy of the supreme source, Viṣṇu or Kṛṣṇa.[1] It must be revered.

Moreover, nature has jurisdiction over all beings. Humans do not have nature's permission to destroy them.[2] People may use nature for life's necessities. But they should not use nature and its creatures for excess pleasure and comfort.[3]

That is logical but not the view of most people in modern society.

When weeds, insects, rodents, or other animals trouble humans, they kill them. In some cases, that is done to preserve life. When something threatens one's life, the *Vedas* declare that one may respond in kind.[4] However, in modern society, most killing is for comfort or to avoid inconvenience.

The clearing of entire forests and mass slaughter of animals is not required to sustain life.[5]

Vegetarianism

In Sanskrit, the word for meat implies that the animal whose flesh I eat now may eat mine in the next life.[6]

Therefore, a non-meat diet is the ideal standard in Vedic culture. However, the harsh truth is that humans always disturb nature in some way. Even a plant diet causes harm to plants. Also, farming upsets nature.

Thus, the *Vedas* assert that society must atone for causing injury to nature, even for the killing of plants and small life forms. Making offerings to the controllers of nature is the best way to compensate for the harm. That compensation is for abuses to nature.[7]

However, just paying a fine for disturbing nature is not enough. People must pay for both use and abuse.[8]

The Transcendental Path

There is also a sublime aspect of making offerings.

They establish an exchange of affection. That love, along with service, is elevating.[9] It counteracts selfishness and brings joy and satisfaction.

The *Vedas* do not just caution against taking life. They prescribe offerings that are atonements, debt payments, and gifts of love. Such offerings are essential in the lives of the best followers of Viṣṇu. And everyone should adopt them.

---

[1] *Īśa Upaniṣad* 6: "One who sees all beings as life force, energy of the supreme source, does not harm anyone."

Bhagavad-gītā 7.5: "The life force that animates all beings is Kṛṣṇa's superior energy. That life force maintains the moving, living universe."

[2] All beings have a role to play in the maintenance of nature and the universe. Humans do not have the right to arbitrarily disturb those roles by killing or dislocating other beings.

Along with all animals, humans tend to act destructively through defense, subsistence, and accident. However, the *Vedas* declare that humans must take care to act responsibly. That means those destructive acts must be done minimally.

[3] *Bhagavad-gītā* 10.5, 13.8, and 16.2 emphasize *ahiṁsā* (nonviolence). 17.4 states that nonviolence is a characteristic of austerity.

Bhāgavata Purāṇa 12.3.30 states that violence is symptomatic of an ignorant society.

[4] *Manu-smṛti* 8.350: "Without consideration, one should kill aggressors, as there is no fault in killing them."

[5] Part of Vedic tradition is the idea of using only dry wood and not cutting down trees or their green branches. Although not always followed strictly, it is considered ideal and more in tune with universal *dharma*.

An example of that mentality was Guru Jambheshwar (1451-1536), who proscribed his followers from cutting green trees and branches. There are currently about 600,000 followers of his sect, mostly in North India. He taught Viṣṇu-centered Vedic monotheism.

[6] *Manu-smṛti* 5.38: "As many hairs as the slain beast has, that many births and deaths will its killer suffer."

Manu-smṛti 5.45: "He who injures innocuous beings from a wish to give himself pleasure never finds happiness, neither living nor dead."

[7] *Bhāgavata Purāṇa* 6.1.8. "During one's life, one should atone by following processes of the *Vedas*. Otherwise, one's time will be lost, and the impending reactions of one's offenses to nature will increase.

"As a disease is treated according to its gravity, one should atone based on the severity of one's offenses."

[8] This concept is clear if one understands that Viṣṇu is the supreme landlord. When we occupy property that belongs to a landlord, we must pay for using it. We also compensate him for damages we might cause.

[9] *Bhagavad-gītā* 9.26: "If one offers Kṛṣṇa even a leaf, fruit or water with love and devotion, He will accept it."

Pleasure

The Greatest Enjoyment

This is the second and most important of the four barriers that guard against overexploitation. It also provides a path to satisfaction that replaces the quest for material comforts.

It is the pleasure of celebrating *līlā* and *mantra*.[1]

Līlā

Vedic culture is *līlā* culture.

Līlā means play, sport, or drama. It refers to the dramas played out by an incarnation of the supreme source. To humans, they appear to be life events. But for Viṣṇu, they are dramas fully under His control. They show another side of the governing source. He is not limited to majesty and justice. He has fun.

And gender does not constrain Him. Half of His expression is as a female.

Līlās involve victory and joy. Kṛṣṇa was victorious

over many opponents. And beyond supreme displays of power, His *līlās* express pleasure, in which there are intimate exchanges with friends, family, and lovers.

The people in Vedic society celebrate *līlās* as a central part of their lives. Instead of overexploiting nature, they enjoy public and private festivals, parades, home worship, temple worship, art, drama, dance, music, sculpture, architecture, science, literature, and much more—all focused on the *līlās*.

Spending money on them redirects it away from the obsession with comforts. Using a portion of society's production to celebrate *līlās* is the way to balance society and overcome the issues with nature.

That is not a theory. It has worked in the Vedic culture for millennia. And it still works in modern Vedic society, even with the materialistic distractions caused by global culture.

The *Līlā* Path

The pleasure offered by *līlās* is much more than a barrier preventing the exploitation of nature. It is a delightful diversion from such exploitation.

It offers a path to the highest pleasure possible, which is direct connection with the source. The source, Kṛṣṇa, is the reservoir of pleasure. Contacting Him is the natural goal of all beings, who are continually searching for enjoyment as part of their existence.

Unfolding the Universe is a *Līlā*

Even the fun of unfurling, maintaining, and withdrawing the physical realms are *līlās* that Viṣṇu enjoys.[2]

Throughout history, He has been revered and depicted in art and sculpture as the causal source of creation.

Moreover, an important aspect of many of His *līlās* is the preservation of *dharma* and the *Vedas*, which are keys to maintaining the universe.[3]

Art and Entertainment

Like normal stories, *līlās* have action, adventure, romance, drama, and comedy. But their unique aspect is that, unlike common stories, they have the power to reduce worldly desires.[4]

The *līlās* entertain the people. And in each region, they have their own flavor.

They are taken from the epics, histories, and ancient writings that are extensions of the *Vedas*. They are statically expressed in media like stone, wood, metal, clay, or canvas.

And the *līlās* are enacted through dramas, dances, puppet shows, songs, and storytelling. There are local and traveling troupes and bards. For thousands of years throughout Asia and beyond, those enactments were the main entertainment before cinema.

Architecture that incorporates *līlā* themes is based on region. And the grand Viṣṇu temples entertain the people. They sponsor huge *līlā* festivals, and people travel long distances to attend.

For ages, pilgrims have crisscrossed continents to visit holy sites, bathe in the sacred rivers, associate with the renounced sages, and attend *līlā* festivals. Such travels are a big part of the culture.

Mantra

A *mantra* is an incantation, a song of praise, a sacrificial formula, or all those together.

Mantras composed of names of Viṣṇu are linked to *līlās*. The many names of Viṣṇu and Kṛṣṇa are often descriptions of the *līlās*. For example, Kṛṣṇa is called

Vaṁśi-dhara, one who plays the flute. Viṣṇu is called Vāmana because He once appeared as a dwarf. And there are thousands of such names that correspond with the thousands of *līlās*.

People sing and chant the *mantras* to celebrate the *līlās* as well as the divine names themselves.

The sages explain that the name of Viṣṇu is absolute. As such, it is the same as His person. Thus, while chanting, He is dancing on the tongue.[5]

And the *mantra* experience is the highest pleasure. It is not selfish. It is pure service. It is uplifting for those who chant and those who hear.

In the Vedic culture, reciting *mantras,* like *om namo bhagavate vāsudevāya, om namo nārāyaṇāya,* and *hare kṛṣṇa,* is an essential element of daily life.[6]

The *hare kṛṣṇa mantra* has three words, *hare, kṛṣṇa,* and *rāma*. It is *hare kṛṣṇa hare kṛṣṇa kṛṣṇa kṛṣṇa hare hare; hare rāma hare rāma rāma rāma hare hare*. Devotees often recite it hundreds and even thousands of times daily.

Mantra and Nature

Mantras prevent the exploitation of nature. On the

one hand, the effects can be directly observed. On the other hand, the influence is subtle or mystical.

Chanting *mantras* detaches one from the desire to consume too much. That one can observe. Those who chant exploit fewer of nature's resources.[7]

The subtle or mystical aspect of chanting is that it invokes a connection with Viṣṇu. Since ancient times, chanting *mantras* has been the way to enjoy a link with the controllers of nature and the supreme controller, Viṣṇu. Becoming attached to Viṣṇu guarantees detachment from exploiting nature.

As the chanting becomes less selfish, the link grows stronger. Still, whether selfish or unselfish, when one recites Viṣṇu *mantras*, the result is good.[8]

The *Vedas* give the analogy of a tree. Watering the root of a tree is the way to nourish all its parts. Thus, offering *mantras* to the root of creation is the way to nourish humanity.[9]

And giving enjoyment to the supreme enjoyer is the best way to enjoy life.[10]

It is also the best way to enjoy life without harming nature. And even beyond such pleasure, the chanting nurtures love of Kṛṣṇa.

Work, Money, and Pleasure

There is a vital link between work, money, and pleasure.[11]

All societies have occupations that produce something, predominantly money. Thus, work links to production or money, and money connects to enjoyment.

The *Vedas* declare that occupations are under the control of nature. Nature gives everyone certain qualities that enable them to pursue an occupation.[12]

In addition, nature controls the results that people receive from working, usually money.[13] Results certainly involve human endeavor, but they are ultimately awarded by nature.

Therefore, people have limited control over their work and its results.[14] But they control how they use the money. Thus, the *Vedas* focus on that area. The process of weaning individuals and society off the selfish use of money is *yajña*.[15]

Vedic *Yajña*

Yajñas are offerings to Viṣṇu or His subordinate

controllers of nature.

People think of offerings as animals or plants placed into a fire or on an altar. But an offering can mean celebrating *līlās,* chanting *mantras,* studying the *Vedas,* or meditation, austerity, and charity in service to Viṣṇu.

The *Vedas* explain that the body is born by the influence of *karma* and *yajña.*[16] Note that despite modern concepts of *karma,* it simply means everyday work.[17]

Karma is the work that people do to obtain money or harvest. And *yajña* is the practice that guides how that money should be used. Thus, *yajña* and *karma* are linked. *Karma* alone without *yajña* is degrading.[18] When properly linked with *yajña,* the *karma* is liberating.

Yajña is based on the concept that the controllers of nature are the true enjoyers, and the ultimate enjoyer is Viṣṇu.[19] When humans recognize their debts to the higher powers and act on that knowledge through *yajña,* society achieves balance with nature.

Enjoyment through *yajña* is the process that elevates individuals and society in this world and the next.[20]

Even if one does not believe in a higher power, one

should still offer the results of one's work in the spirit of *yajña* to benefit society and nature.[21]

Seeking pleasure through *yajña* controls the desire to exploit beyond one's quota. That is because it reduces attachment caused by selfish desire.

People in modern society use money for increased comfort. The obsession to consume is born and grows from attachment to those comforts.[22]

In contrast, people in the Vedic culture value detachment from selfish desires. *Yajña* is one of the ways they achieve it.[23]

And the celebration of Viṣṇu's *līlās* and chanting of *mantras* are the greatest *yajñas*.[24]

---

[1] A *mantra* is an incantation, a song of praise, sacrificial formula, or all those together. *Mantras* composed of names of Viṣṇu are linked to *līlās*.

[2] *Bhāgavata Purāṇa* 3.26.4. "As His *līlā*, the greatest of the great Viṣṇu, unfurled nature from subtle to gross."

[3] *Bhagavad-gītā* 4.7: "Whenever there is a decline in *dharma*, I appear on earth."

Gītā Govinda, Daśāvatāra Stotra by Jayadeva (c 1200). This is a beautiful song that summarizes ten of Viṣṇu's *līlās*.

[4] *Bhāgavata Purāṇa* 10.33.39: "One who hears or describes Viṣṇu's (Kṛṣṇa's) *līlās* with conviction achieves the highest service to Him. That wise person quickly drives the disease of worldly desire from his heart."

---

[5] *Stava-Mala* by Sri Rupa Gosvami: "The *mahā-mantra* ecstatically dances on the tongue"

[6] *Kali-Saṇṭāraṇa Upaniṣad* 5-6: "*Hare kṛṣṇa hare kṛṣṇa kṛṣṇa kṛṣṇa hare hare hare rāma hare rāma rāma rāma hare hare*--these sixteen names are the only means to counteract the evil effects of Kali-yuga (the current dark age)."

Bhāgavata Purāṇa 1.1.1: *om namo bhagavate vāsudevāya*, "Adore that supreme source Vāsudeva, Kṛṣṇa."

Bhāgavata Purāṇa 6.5.28: *oṁ namo nārāyaṇāya*, "Prostration to Nārāyaṇa, Viṣṇu."

Bhāgavata Purāṇa 12.3.52: "Whatever result was obtained in Satya-yuga by meditating on Viṣṇu, in Tretā-yuga by performing sacrifices, and in Dvāpara-yuga by serving the Lord's lotus feet can be obtained in Kali-yuga simply by chanting the Hare Kṛṣṇa *mahā-mantra*."

Śrī Caitanya-caritāmṛta Ādi-līlā 17.21: "In this age, there is no other way, none other, indeed no other path than the name of Hari, who is the supreme source."

[7] *Bhagavad-gītā* 2.59: "By experiencing enjoyment in connection with transcendent reality, one can give up the taste for physical enjoyment."

Bhagavad-gītā 6.5: "The mind can be friend or foe. It should be used to elevate not degrade."

Padma Purāṇa, brahma-khaṇḍa 25.15–18: "To remain attached to physical reality is offensive to the name."

[8] *Kali-Saṇṭāraṇa Upaniṣad* 9: "There are no rules or procedures for chanting. One may be pure or impure. The result is the same: connection to the supreme source."

[9] *Bhāgavata Purāṇa* 4.31.14: "As watering the root of a tree or feeding the stomach is the proper way to nourish, so serving the source of everything is the way to nourish all beings."

[10] *Bhagavad-gītā* 5.29: "Kṛṣṇa is the enjoyer of all sacrifice (*yajña*)."

This means that *yajña* establishes a link with Kṛṣṇa based on enjoyment. The link acknowledges that Kṛṣṇa is the supreme enjoyer. And we enjoy by giving Him enjoyment, like watering the root of a tree.

[11] *Bhagavad-gītā*, Chapter 3.

[12] *Bhagavad-gītā* 18.41: One's profession is indicated by the qualities that nature awards at birth.

[13] *Bhagavad-gītā* 5.14: "Nature awards the results of one's work."

[14] "Limited control over their work," means that people cannot control what qualities nature assigned them at birth. Their qualities constrain them to a certain type of work. And their qualities enable them to excel at something.

[15] *Bhagavad-gītā* 3.13: "The wise use accumulated natural resources for *yajña*. Those who consume them for selfish purposes cause suffering."

[16] *Bhāgavata-Purāṇa* 1.13.46: "The body is under the control of time, *karma*, and nature."

Bhagavad-gītā 3.10: "Humans are created along with *yajñas*."

[17] The Sanskrit dictionary defines *karma* as an act, action, work, occupation, etc. It is from the verb root, *kṛ*, to do, make, perform, etc.

[18] *Bhagavad-gītā* 3.9: "Work (*karma*) must be done as a *yajña* for Viṣṇu. Otherwise, *karma* causes bondage to suffering."

[19] *Bhagavad-gītā* 5.29, cited above.

[20] *Elevation in this world* means elevation to the level of illumination and liberation from the influence of *māyā*.

[21] *Bhagavad-gītā* 12.11: "Even if you cannot bring yourself to offer to a higher power, still, you should give up the results of your work (the money) in *yajña* and be self-satisfied."

Here, the principle is that at least one should become detached from money by using it for a higher purpose. In other words, using money for personal and family life beyond the necessities for care and maintenance is not recommended. It causes suffering and bondage to the physical realm.

[22] *Bhagavad-gītā* 2.62.

[23] The practices of *bhakti-yoga,* which award detachment from selfish desires is also a form of *yajña.*

[24] *Bhāgavata Purāṇa* 11.5.37: "There is no greater gain for humanity than celebrating Viṣṇu's *līlās* and names (*saṅkīrtana*). By doing so, one attains liberation from the cycle of birth and death."

Agrarianism

An agrarian economy is the third of the four barriers that help prevent imbalance with nature.

Vedic cultural ecology defines agrarianism as natural farming and cottage industry. Production and consumption are mostly local.

It is the traditional economy that existed for thousands of years before colonialism, agribusiness, and heavy industry.

Through colonialism, Europeans eroded rural, locally based farming and trade. They sought more resources through worldwide commerce and conquest. And they developed mass production in farming.

Now, the traditional economies once everywhere in the world are nearly extinct. Many doubt there can be a return to the simple, local farming economies.

But most people in the Vedic culture place a high value on rural life and the economy that goes with it. It requires more discipline than city life, but it is essential for a healthy society.

Agrarianism must be renewed.

Rural versus Urban

Throughout history, large cities and city-states have overused their local resources.[1] Now that is happening to much of the earth. Cities that destroy nature have expanded like plagues and enveloped nations.

Humans naturally establish centers of art, trade, and government. Those exist alongside flourishing markets, which encourage the pursuit of comfort. And people think that their urban centers are great achievements. That has been the mentality for ages.

Now the attitude in modern societies is that people are more advanced than in any past culture. Despite their pride in being modern, people still cling to the same old primitive idea that they have a right to overuse nature, which is a dead-end mentality.

The values of simple, local farming must be adopted. Cities are indeed essential as centers of art, trade, and government;[2] however, when the only purpose of rural areas is to serve the cities, an unnatural imbalance is created.

Widespread vibrant agrarian life must balance a few, more exploitative cities.

A rural economy based on farm produce, cottage industry, and service to the transcendent source will sustain humans for thousands of years.

That is the natural economy of the Vedic culture. It does not reject cities, but instead maintains a balance between urban and rural.

Social Structure

Social classes exist in both Vedic culture and modern global society. In Vedic culture, they correspond to human nature.[3] They are designed to facilitate two goals: sustaining life and progressing toward liberation from suffering due to the identification with physical reality.[4]

In modern culture, the classes are broadly based on wealth and power.[5] They develop from the desire to exploit natural resources for greater comfort. There is very little guidance toward freedom from attachment to those desires. Instead, the entire social structure—its education and occupations—is tied to those desires. Thus, modern society is not based on freedom. It is in bondage to desire and greed, which cause suffering.[6]

The classes in Vedic culture are guided toward service, satisfaction, and illumination. On the surface, there is inequality. But in pursuit of the goals, which are life and liberation, there is equality.

In modern society, there is a constant push for greater equality. However, the destruction of nature guarantees inequality as people fight for resources.

The Cycle of Life

The life cycle is birth, maturation, decline, and death followed by rebirth, which continues the cycle.

The direction of one's life is influenced by participation in the *yajña*-cycle. Those who participate are awarded a body with more opportunity to serve the supreme source.

Those who do not work for *yajña* receive a body covered by a cloud of greater selfishness or ignorance.

The simple agrarian communities, along with the natural Vedic social structure, are designed to facilitate participation in that *yajña*-cycle. That is the higher path offered by Vedic agrarianism.

---

[1] "Overexploitation of Renewable Resources by Ancient Societies and

the Role of Sunk-Cost Effects," by Marco Janssen and Marten Scheffer, *Ecology and Society* 9.1

[2] Here *art* includes the arts, sciences, philosophies, etc. These are aspects of a sophisticated or intellectually developed culture.

[3] *Bhagavad-gītā* 18.41: "Society is organized into four classes based on natural characteristics that people have at birth. Those characteristics qualify one for a type of work."

[4] Becoming free from the desire to exploit includes liberation from physical attachments or the desire for physical enjoyment.

There is a third goal of Vedic culture. That is to connect with the Supreme source, Viṣṇu or Kṛṣṇa.

However, the two goals, sustenance and liberation, correspond to *varṇa* (social class) and *āśrama* (stage of life). The higher goal of linking with Viṣṇu pervades both in an ideal (*daivic*) society.

[5] People are categorized as lower, middle, and upper class. They are further divided into sub-classes. The basis of those classes is broadly wealth, power, and sometimes privileged birth.

[6] The *Vedas* do not recommend becoming free from all desires, only the desires that promote attachment to the physical realm.

Those desires should be replaced by the intense desire to reach transcendent reality and connect with the supreme source, Viṣṇu.

Tolerance of Nature

Tolerance

This is the fourth of the four barriers that control the overexploitation of nature in Vedic culture.

The *Vedas* teach that one must tolerate nature. That means the hardships of winter, summer, old age, and pain. They are beyond our control.[1]

In contrast, a major theme in global culture is to fight those problems. The cold of winter is met by burning fossil fuels. Air conditioning, also powered by fossil fuels, reduces the heat of summer. And everything possible is done to eradicate pain.

It is natural for some people to have a weak body or mind and seek comfort away from hardship. But too many people in the global culture are entitled, weak, and addicted to comfort compared to natural society.

Those in Vedic society know that enjoyment and suffering do not last. They come and go.

Too much attachment to physical enjoyment and

aversion to suffering cause imbalance in society. Tolerating the inevitable changes that affect the body and mind is the answer.[2]

Granted, tolerance is not possible for everyone. But through austerity and discipline, one may achieve it.[3]

Austerity

Austerity is an essential part of Vedic culture. In modern society, people discipline themselves to develop their bodies, study, or work in a profession.

But in Vedic culture, austerity means tolerating nature. That includes overcoming harmful and selfish qualities that are part of human nature.

Humans tend to seek too many selfish comforts. And the *Vedas* offer prescriptions to counter that tendency. They structure Vedic society according to four stages of life.[4]

First, young people are trained to enjoy self-discipline and tolerance of nature.

Second, in adulthood, those who are building a family are expected to care for it using more comfort than in the other stages. Thus, that is the least austere stage. But families use their money for *yajña*, not just comfort.[5] That is their form of austerity.

In the third and fourth stages in old age, greater tolerance of nature is again cultivated.[6]

And none of that can be done without learning the joy connected to austerity.

The *Vedas* describe austerity in accordance with the levels of selfishness.[7] The lowest causes harm.[8] Austerity for personal gain is on the intermediate level.[9]

At the illuminated level, one's mental, physical, and verbal behaviors are the least selfish.[10]

The followers of Viṣṇu take seriously certain basic austerities. They reject eating meat, having sex outside marriage, taking drugs or alcohol, and gambling. And they use their money for *yajña*.[11]

Certainly, not all the billion people in Vedic society are at the highest standard. Still, most have great respect and deference for those values.

Renunciation

Society is out of balance when most people live in cities. The proper distribution of the population is that most should live in rural areas.

Moreover, a few may live in places removed from the comforts and passions of cities, towns, and villages. The *Vedas* urge those obsessed with consumption to go away from the places of temptation to balance their appetites.[12]

One of the most prominent kings in history, King Bharata, gave up his throne and retired to the forest. South Asia, which is called Bhārata (or Bhārat) in the native languages, was named after that king.[13]

That shows the respect people in Vedic culture have for renunciation. Although ruling an empire requires a majestic life, an important goal of life is detachment from *māyā,* which may be achieved through austerity.

Currently, it is not practical for people to retire to the forest. Instead, one may live in or near a Kṛṣṇa temple to relieve oneself of the obsession with consumption.

Kṛṣṇa Culture

Modern global culture is far removed from such austerities. But a simple life without too much consumption is possible. That can be done by cultivating the proper type of enjoyment. Doing so may require living in a place away from the addiction to consumption. It is a good idea for retired people to live in a Kṛṣṇa community near a temple or farm and spend most of their time serving there.

And before beginning a family, young people should live in a temple and learn the joys of monastic austerity.

Aside from those disciplines, modern global culture must change its heroes. Its current idols—sportspeople, actors, and musicians—are examples of excess comfort and consumption.

The new heroes must be those who live simply and dedicate their lives to the service of Viṣṇu and Kṛṣṇa. Instead of overconsuming, those heroes are committed to educating the public about the wisdom of the *Vedas*.

The Path of Tolerance

When joined with the other paths mentioned in this section, the path of tolerance guides society to transcendent reality.

All four paths award true satisfaction.

In contrast, people under the influence of *māyā* seek physical comfort, often through overconsumption. Physical enjoyment masks pain and suffering. Some become skilled at such masking, especially through wealth and power.

However, at minimum, old age, disease, and death are signals that one should not traverse the path of *māyā*.

Instead, the four Vedic barriers to greed and paths to liberation should be adopted by modern society for the benefit of all.

---

[1] *Bhagavad-gītā* 2.14.

When a cultured person in Vedic society reaches retirement age, he is urged to give up comforts and tolerate nature.

[2] The reader should not consider this section to be a personal attack against his or her lack of tolerance or austerity. The arguments here are aimed at modern society, which is based on the pursuit of luxury at the expense of nature.

Vedic society has social structures and disciplines in place that combat the natural human tendency for excess comfort. They cannot be implemented overnight.

Thus, the reader is urged to understand the whole Vedic culture. The parts of it that encourage austerity and renunciation are best developed organically and gradually.

[3] *Manu-smṛti* 11.239: "Whatever is difficult to be practiced, endured, or attained may be accomplished by austerity, for it possesses a power difficult to surpass."

[4] *Bhāgavata Purāṇa* 11.17-18 and 7.12-15 describe the stages of life (*āśramas*). Those stages are meant to gradually elevate one to liberation

from identification with physical reality.

Not everyone is expected to observe all stages; however, the sophisticated or genteel parts of society are encouraged to do so as much as possible.

[5] *Bhāgavata Purāṇa* 7.14.15: "Every day, with the money he [the householder] has acquired, he should perform *yajña* to the Supreme and separately to the *devas*, *ṛṣis*, humans, *pitṛs*, other beings, and himself."

Families also provide charity to other beings, the less fortunate, and those in the other stages of life. Three essential aspects of *dharma* are *yajña*, charity, and austerity. (*Bhagavad-gītā* 18.5)

[6] *Bhāgavata Purāṇa* 11.17-18 and 7.12-15—chapters on the stages of life.

[7] *Bhagavad-gītā* 17.7: "There are three levels of food, *yajña*, austerity, and charity."

[8] *Bhagavad-gītā* 17.16: "The lowest austerity is foolish and ignorant. It is done to injure."

[9] *Bhagavad-gītā* 17.18: "Selfish austerity is motivated by pride, ambition, and personal gain. It is not lasting.

[10] *Bhagavad-gītā* 17.14-16: "Austerity of the body includes proper honor to the Supreme and His devotees, purity, simplicity, celibacy, and nonviolence.

"Austerity of speech is truthful, pleasing, beneficial, not provocative, and based on the *Vedas*.

"Mental austerity is satisfaction, simplicity, gravity, self-control, and purification."

[11] *Bhāgavata Purāṇa* 1.17.38-39: "Kali, the friend of *adharma*, was compelled to remain where there is gambling, illicit connection with women, animal slaughter, and intoxication.

"He was also assigned to money that is used for overconsumption, addiction, selfishness, and hate."

[12] *Bhāgavata Purāṇa* 7.5.5: "One who is completely bound to personal and familial responsibilities, which cause identification with physical reali-

ty, should go to the forest and seek to connect with the Supreme Viṣṇu."

[13] *Bhāgavata Purāṇa* 5.7-13: These chapters relate the story of three lives of King Bharata.

India is called Bhārat in the native languages. The name for South Asia is Akhaṇḍ (undivided) Bhārat, meaning the whole of South Asia undivided into separate countries.

Bhagavad-Gītā

The *Gītā* is a summary of the core values, concepts, and wisdom of the *Vedas.*[1] As such, it is an extension of the *Vedas.*[2]

It explains that one achieves true joy by detaching oneself from the desires of the body and things related to it. The purpose of doing so is to achieve contact with the life force and the supreme source, Kṛṣṇa, which together constitute transcendent reality.[3]

The method for connecting with transcendent reality is *yoga.*[4] And *bhakti yoga* is the most practical and direct approach.[5]

It starts with the mind. The mind should control the senses. But instead, the senses have taken control of it.[6]

Yoga is the means to regain control of the mind.[7]

The mind falsely serves the senses, which are attached to physical objects.[8] *Yoga* redirects the mind, severing that attachment and linking it to the Supreme.[9]

The highest *yoga,* which is *bhakti,* redirects a person

from service to the senses to the service of Kṛṣṇa. It introduces a form of enjoyment that transcends physical pleasure.[10]

Yajña is intertwined with *yoga*. It is designed to reduce attachment to worldly objects, especially money.[11] Using money, speech, intelligence, and life's endeavors to serve Viṣṇu is *yajña*.

In modern society, it can bring about a revolution.

Adopting *yajña* will change society's relations with nature. And doing so is not an unproven idea. Hundreds of millions in Vedic culture show how it can be done.

Societies are always changing. All societies are fluid. Thus, there is no reason for modern global society to refuse to adopt the principles of Vedic society.

The concepts and principles—that is, the *dharma*—of *yoga* and *yajña* are presented in the *Gītā*. And they are applied in the Vedic culture. Any person who wishes to become truly wise must study the *Gītā*, experience the Vedic culture, and attempt to implement it in modern society.

---

[1] *Bhagavad-gītā* 9.2: "This is the king of knowledge, secret of secrets, and

most pure. Its teachings may be directly perceived. It is the perfection of *dharma*. Applying the teachings produces enjoyment and satisfaction."

[2] *Gītā-māhātmya* 6: "The *Gītā* is the essence of the *Upaniṣads*."

[3] *Bhagavad-gītā* 3.43: "Knowing oneself to be above the physical reality, one should steady the mind, and with proper discrimination, conquer physical attachment and desire."

[4] *Bhagavad-gītā* 6.28: "The *yogī* who is disciplined and focused on the self (the life force) becomes free from the physical world and achieves perfect happiness, being linked to the supreme source, Kṛṣṇa."

In previous chapters, following *dharma* was recommended to change from identification with the body to the life force. Here it is *yoga*.

That is because universal *dharma* does indeed provide a gradual path of elevation to transcendent reality.

However, the supreme *dharma*, which is beyond universal *dharma*, is the *yoga* that directly causes one to identify with the life force and connect with the supreme source, Kṛṣṇa.

That supreme *dharma* or *yoga* is taught in the *Bhagavad-gītā* and *Bhāgavata Purāṇa*.

[5] *Bhagavad-gītā* 6.47: "Of all *yogīs*, the one who, with conviction, offers *bhakti* to Kṛṣṇa and is absorbed in Him, is the highest."

[6] *Bhagavad-gītā* 6.6: "The mind that is under control is a friend, while the uncontrolled mind is an enemy."

[7] *Bhagavad-gītā* 2.50: "*Yoga* is the perfection of *karma* (work)."

[8] *Bhagavad-gītā* 2.62: "One who contemplates the objects of the senses (sights, sounds, tastes, etc.) becomes attached to them. From that attachment arises the desire to possess and connect with them."

[9] *Bhagavad-gītā* 2.64: "One who is free from attachment and aversion and restrains the senses from contact with the objects achieves the goal (liberation)."

Bhagavad-gītā 15.3: "With determination, one must sever identification with physical reality by the weapon of detachment."

[10] *Bhagavad-gītā* 2.59: "By experiencing transcendental enjoyment, one can give up the taste for physical enjoyment."

[11] *Bhagavad-gītā* 3.9: "One's *karma* must be part of *yajña* for Viṣṇu."

Bhagavad-gītā 3.10: "At the beginning of creation, Viṣṇu arranged for *yajña* so that humanity might live happily."

The Next Step

Much of this book has been critical of modern culture, including its fundamental institutions like modern science and democracy. However, the purpose is not to eliminate them.

Although cultural change is recommended, the highest principle is maintenance. Making changes must be made properly.

Revolution by violence is ignorant.

Also, creating a new government or culture without properly establishing *dharma* is inadequate.

Instead, to properly maintain balance with nature, a section of society, roughly one to five percent, who can steer the majority in the proper direction must be nurtured.

That section—those people—should be detached from wealth and possessions. They should strongly identify with transcendent reality and minimally with physical reality. They should be exemplars of unselfish service to Viṣṇu.

Such a group must be cultivated, and with the

support of government and the business community, they must be empowered to influence change.

When detachment from the physical realm becomes more important than power, fame, and wealth, society will have begun its journey toward balance.

Humans are naturally greedy. To balance it, those among us with the least amount of greed must be the most respected and exalted.

The primary symptom of such people is that they do not desire power or wealth for their selfish purposes, meaning for themselves and their families.

Nevertheless, they must be given influence over those who do have power and wealth.[1] In that way, they can guide society toward service to humanity, nature, and the supreme source of nature, Kṛṣṇa.

To do that, they must be intelligent in the Vedic sense. They must have *buddhi* or the power to discriminate between *dharma* and *adharma*—what should be done, how power should be applied, and how wealth should be employed.

Now, please send an email for information about a Zoom call to info@cvc.guru and join us for the next step.

[1] "Influence of the proper direction" means such illuminated people do not hold positions of wealth or power. Instead, they have strong influence over those who do.

Appendix

For info on Zoom meetings, email info@cvc.guru.

Terminology

This chapter is for devotees of the *Vedas, Gītā,* and *Purāṇas*. They may be unaccustomed to some of the terminology used throughout the book.

Here is our justification for using various terms.

Material and Physical

Throughout the book, *physical* is the preferred term.

Material, as in material nature, used in many translations of the *Gītā, Purāṇas,* etc., has been replaced here by *physical*.

In modern vernacular, physical tends to encompass more than material. It is also used by scientists, whereas material tends to be used by philosophers and religious people.

Physical Reality

Physical reality is used here for *prakṛti* or material nature. Reality is *tattva* in Sanskrit.[1]

Causal or Governing Source

The combination of *prakṛti* and *puruṣa* causes the

universe to unfold.

With the glance of time, *puruṣa* agitates *pradhāna,* thus transforming it into *mahat-tattva,* which then unfolds into the material or physical elements.[2]

That *puruṣa* is Viṣṇu.

Kāraṇa-udaka-śāyī Viṣṇu lies in the Kāraṇa (causal) Ocean. He is the ultimate cause or source of innumerable physical realms or universes.

Kṣīra-udaka-śāyī Viṣṇu is the third expansion. He lies in the Milk Ocean. He is the source of all incarnations. He maintains the universe, and He pervades it as Paramātmā.[3] He plays the part of the governing force or maintainer of the universe.

Source indicates the ultimate cause or cause of all causes—Kāraṇa-udaka-śāyī.

Governing means that the causal source also maintains the universe as Kṣīra-udaka-śāyī.

The *Vedas* are clear that the universe does not have a random source. It is not probabilistic; it is deterministic; it is governed.

Universal *Dharma*

Universal *dharma* and nature's *dharma* are the terms used here for *sanātana dharma* or *varṇāśrama dharma.*

Descriptions of that *dharma* are in the *Bhāgavatam* 7.13-15 and 11.17-18.

God

In Sanskrit and other South Asian languages, God is generally translated as Bhagavān.

However, the *Bhāgavatam* states that those who know reality (*tattva-vidas*) understand that *tattva* (the supreme reality) is threefold: Brahman, Paramātmā, and Bhagavān.

Therefore, the word God is inadequate. Its use may lead to misunderstanding.[4] Vedic cultural ecology minimizes the use of that word.

Spiritual

Spiritual is *ādhyātmika* or that which is related to *ātmā*. It can also be *brahmaṇya,* which means related to *brahma* or spirit.[5]

In English, the old religious use of spiritual generally meant that which relates to the Holy Spirit. That use is roughly analogous to *brahmaṇya*.

However, in modern vernacular, spiritual is often used by those who do not wish to identify with religion.

It can mean mystical, cultural, or artistic.

It may also mean something that produces significant emotion: "We had a spiritual connection," which may even be sexual or sensual.

Thus, due to the modern misuse of spiritual, Vedic cultural ecology avoids it.

Instead, the preferred term here is *transcendent reality* or *Vedic culture*.

Life Force

Instead of soul, *life force* is used here. The word *soul* is used in Christianity, and it typically means something different than *ātmā* or *jīvātmā*.

Life force implies that something inanimate (the body) is animated by that force (*jīvātmā*).[6]

Conceptions of the Source

In the *Vedas*, there are a few concepts of the supreme source.[7]

In the first sections of this book, the term *source* is broadly used to indicate that one may differ about the definition of the source but not its existence. In other words, atheism is unacceptable. It is illogical and socially irresponsible to deny a source of the universe.

The section on Vedic Culture establishes Vedic

monotheism as the best.[8] Again, cultural arguments predominate rather than philosophical ones.[9]

The argument behind the above progression is that atheism, prevalent in modern culture, is abhorrent. It is ruining society.

Moreover, nominal faith in God is not enough. If society is overexploiting Kṛṣṇa's nature, it is fundamentally an atheistic society.

Instead, a society must be source-centric (Kṛṣṇa-centered), and the citizens must be directed toward liberation from material enjoyment and bodily identification.

Vedic Pantheism

Brahman realization; Brahman as transcendent reality.[10]

Vedic Monotheism

Bhagavān worship—*bhakti* to Bhagavān Śrī Kṛṣṇa.[11] Bhagavān is the highest form of transcendent reality.

Vedic Polytheism

Demigod worship. Worship of the *devas*. From Greek, *poly* (many) *theos* (god).

---

[1] *Tattva* is used in the term *mahat-tattva*, meaning the total material energy.

[2] *Puruṣa* means person. In describing creation, He is the Supreme Person. *Pradhāna* is the unmanifested fountainhead of nature.

[3] *Śrīmad Bhāgavatam* 3.7.22 purport by A.C. Bhaktivedānta Swami, Śrīla Prabhupāda.

[4] The word *God* has a disputed etymology. It is generally considered to be a pre-Christian word from German via Old English. It may have originally meant a deity that is invoked for a sacrifice. The word was later adopted by Christianity.

[5] *Brahmaṇya* also means devoted to the *brāhmaṇas*.

Brahminical (Vedic) culture is spiritual because *brāhmaṇas*, who are the leaders of the culture, are devotees of *brahma*, the Supreme Spirit.

[6] *Bhagavad-gītā* 7.5.

[7] *Bhagavad-gītā* 9.15–18 lecture by A.C. Bhaktivedānta Swami Śrīla Prabhupāda, Dec. 2, 1966: "So this *Bhagavad-gītā* says that those who are trying to make a show of their knowledge, so let them do that. *Viśvato-mukham*. The universal form, pantheism, monotheism, monism—we have so many theories. But not atheism."

[8] This is done in the chapter entitled, "The Supreme Source."

[9] People in modern culture are not moved by philosophical or *śāstric* arguments. They must have empirical (observational) proofs.

Also, without realizing it, their logic is corrupted by emotional attachment to physical reality (sense gratification). That is called *māyayāpahṛta-jñāna*.

Thus, the first task is providing observable proof. That is done by presenting the observation and comparison of Vedic and modern cultures.

The second task is to present arguments that attack emotional attachment. Here, *līlā* enjoyment is used as the weapon to sever that attachment.

"This *hari-saṅkīrtana* is the *astra* (weapon)." [Prabhupāda, 5/29/1976]

[10] *Perfect Questions, Perfect Answers*, by Śrīla Prabhupāda: "The pantheists say that because everything is God, whatever we do is God worship."

Wikipedia, "Pantheism:" Advaita Vedānta is considered pantheistic. There, Brahman alone is reality, and the physical universe is an illusory appearance of Brahman.

[11] *Śrīmad Bhāgavatam* 2.1.37 purport by Śrīla Prabhupāda: "Monotheism is practically suggested here."

Printed in Great Britain
by Amazon

60218040R00099